Web 全栈项目
开发入门与实战

吴杏平 曹雪 / 著

人民邮电出版社

北京

图书在版编目（CIP）数据

Web全栈项目开发入门与实战 / 吴杏平，曹雪著. --北京：人民邮电出版社，2020.9（2024.3重印）
ISBN 978-7-115-54139-0

Ⅰ．①W… Ⅱ．①吴… ②曹… Ⅲ．①网页制作工具—程序设计 Ⅳ．①TP393.092.2

中国版本图书馆CIP数据核字(2020)第093601号

- ◆ 著　　吴杏平　曹 雪
  责任编辑　赵 轩
  责任印制　王 郁　马振武
- ◆ 人民邮电出版社出版发行　北京市丰台区成寿寺路11号
  邮编　100164　电子邮件　315@ptpress.com.cn
  网址　https://www.ptpress.com.cn
  天津画中画印刷有限公司印刷
- ◆ 开本：787×1092　1/16
  印张：32.5　　　　　　　　2020年9月第1版
  字数：619千字　　　　　　2024年3月天津第10次印刷

定价：79.00元

读者服务热线：(010)81055410　印装质量热线：(010)81055316
反盗版热线：(010)81055315
广告经营许可证：京东市监广登字 20170147 号

前言
Foreword

很多大学并没有专门开设针对企业应用开发的项目实训课程。即便是某些计算机专业科班出身的学生，虽然计算机基础功底十分扎实，也很有可能没有独立做过一个完整的项目。正是由于传统的学校没有开设针对企业应用开发的项目实训课程，加之互联网行业是一个高薪行业，因此越来越多的非科班生、转行者涌入互联网浪潮。与之纷至沓来的，是雨后春笋般的各种培训班。但是许多人即便是从培训班学完之后，仍与准备找工作的科班应届生面临同样的问题，即如何能够自主开发一个真实的企业级项目？

随着互联网行业的迅猛发展，用户对 Web 应用要求越来越高，很多全栈工程师愈感力不从心。他们当中有的曾经是前端开发工程师，可以迅速开发出既简洁又美观的界面，实现各种各样的交互效果，但是对后端的数据逻辑却"云里雾里"，不知所以。他们当中有的曾经是后端开发工程师，可以写出效率极高的 SQL 语句，甚至可以快速处理高并发，但是有时候面对最基础的 CSS 样式，面对错综复杂的浏览器兼容性，他们也束手无策，无可奈何。有没有一个完整的全栈（前、后端）项目可以借鉴学习？

越来越多的创业者将目光投向互联网行业。即便不是互联网行业，实业的营销方式也一定离不开互联网的"繁殖性传播"。对于开发一个 Web 应用，不管是作为承载信息的平台，还是作为商品的主营渠道，都显得尤为重要。而大部分创业者都会面临资金紧张、人手不足的情况。如果有能力独自开发出一个应用，岂不美哉？

本书将打造一个完整、经典的全栈（前、后端）实战项目。学完本书后，您收获的不仅是一个项目案例和实战技术，更重要的是，您会了解开发一个完整的项目需要经历哪些阶段，一个完整的项目包含的内容有哪些，在开发项目前如何去构思、分析才能事半功倍。本书不仅"授之以鱼"，更"授之以渔"。

❑ **本书读者对象**

想要开发一个完整的项目，或用于实战演练，或用于求职跳槽的读者。

想要了解一个完整的应用该如何去构思、开发、部署的读者。

想要独立开发一个 Web 应用的读者。

❑ **本书特色**

本书不仅介绍了前后端开发的技术，更融入了许多开发前对业务场景、技术选择的分析过程。比如开发一个页面，不仅会从技术层面讲解如何去实现，更会讲解开发一个页面的分析过程。这种分析的思维几乎适用于所有的页面的开发。

为了保证经验尚浅的开发新手可以举一反三，本书中的项目案例非常灵活，其中涉及的内容都是 Web 应用项目中一些常见、经典的场景。也就是说，如果读者想以后独自开发其他业务类型的项目，只需要在本项目的基础上进行调整即可。比如本书中的项目分为五大业务模块，分别是用户模块、商品模块、支付模块、消息模块及个人中心模块，放眼看去，您会发现大部分的应用都会包含这几个业务模块。再细化到功能，比如项目中的登录、注册功能，也是几乎所有项目都会涉及的功能。

另外，本书介绍的前端开发使用的是 React 框架，该框架最显著的特点就是组件化，因此项目中有大量的组件封装。这就意味着，您将得到的不仅是一个项目，还将得到前端应用中许多场景的通用组件。这些通用组件，读者可以运用到任何其他项目。虽然市面上也有很多成熟组件库，比如蚂蚁金服的 AntDesign，但是对于大部分人来说，这已然是一个"开箱即用"的成品，无法知晓其构建过程。学习了本书之后，读者可以自己构建通用组件。

❑ 资源

本书涉及的资源有源代码、UI 设计图、项目中所用到的图片和图标、业务流程图、业务功能清单表、接口文档等，读者可通过邮箱 wxp67655049@126.com 获取。

❑ 致谢

从小我就有一个当作家的梦想，后来因为种种原因踏入 IT 领域之后便"越陷越深"。能出版 IT 领域的书籍，既是不误现在，亦是不忘初心。在此特别感谢人民邮电出版社的编辑赵轩先生给予我的帮助和支持，也谢谢读者朋友们，是你们一起帮我圆了此梦！

目录

项目概述

第1章 项目需求设计..................02
- 1.1 项目背景..................02
- 1.2 项目介绍..................02

第2章 业务模块介绍..................03
- 2.1 用户模块..................03
  - 2.1.1 用户模块功能清单......04
  - 2.1.2 用户模块功能说明......04
- 2.2 商品模块..................05
  - 2.2.1 商品模块功能清单......05
  - 2.2.2 商品模块功能说明......06
- 2.3 支付模块..................07
  - 2.3.1 支付模块功能清单......07
  - 2.3.2 支付模块功能说明......07
- 2.4 消息模块..................08
  - 2.4.1 消息模块功能清单......08
  - 2.4.2 消息模块功能说明......08
- 2.5 个人中心模块..................08
  - 2.5.1 个人中心模块功能清单..................09
  - 2.5.2 个人中心模块功能说明..................10

第3章 业务流程图..................12
- 3.1 注册页面流程图..................12
- 3.2 登录页面流程图..................13
- 3.3 商品发布页面流程图..................13
- 3.4 首页流程图..................14
- 3.5 消息页面流程图..................15
- 3.6 个人中心页面流程图..................15

第4章 项目开发周期...17

前端开发

第5章 HTML、CSS、JavaScript基础..................22
- 5.1 网页的骨架——HTML..................22
  - 5.1.1 HTML 结构..................23
  - 5.1.2 HTML 头部标签..................24
  - 5.1.3 HTML 内容标签..................25
  - 5.1.4 HTML 样式标签..................27
  - 5.1.5 HTML 脚本标签..................28

5.2 网页的外衣——CSS 29
　5.2.1　CSS 历史...................... 29
　5.2.2　CSS 样式...................... 30
　5.2.3　CSS 选择器.................. 32
　5.2.4　CSS 伪类与伪元素 36
　5.2.5　CSS 使用方式 40
5.3 网页的交互——JavaScript ... 43
　5.3.1　JavaScript 历史............. 43
　5.3.2　JavaScript 语法............. 45
　5.3.3　AJAX 介绍.................... 51
5.4 Web 网页案例...................... 53
　5.4.1　案例说明...................... 53
　5.4.2　案例代码...................... 53

第 6 章　前端主流框架——React 60

6.1 React 概述........................... 60
　6.1.1　React 主要特点 60
　6.1.2　React 生命周期 61
6.2 React 开发环境搭建 62
　6.2.1　Node 安装..................... 62
　6.2.2　React 项目构建 64
6.3 React 案例........................... 67
　6.3.1　案例代码...................... 67
　6.3.2　React 框架与原生网页的区别...................... 69

第 7 章　前端常用开发工具/库 71

7.1 打包工具之 webpack........... 71
　7.1.1　webpack 核心原理......... 71
　7.1.2　webpack 核心概念......... 72

　7.1.3　webpack 参数配置说明 ... 72
7.2 页面跳转之 react-router-dom 75
　7.2.1　react-router-dom 路由配置............................. 75
　7.2.2　react-router-dom 路由跳转方式............................. 76
　7.2.3　react-router-dom 路由传参............................. 76
7.3 前端组件库之 ant-design 77
7.4 CSS 预处理器之 Less 78
　7.4.1　Less 特征...................... 78
　7.4.2　Less 使用环境............... 79
　7.4.3　Less 语法...................... 79
7.5 第三方模块安装................... 81

第 8 章　前端开发前须知 ... 83

8.1 命名规则............................. 83
　8.1.1　模块命名规则............... 83
　8.1.2　选择器命名规则........... 83
8.2 公共样式提取..................... 84
　8.2.1　公共样式规则............... 84
　8.2.2　公共样式文件............... 84
8.3 公共组件封装..................... 92
　8.3.1　Button 组件.................. 93
　8.3.2　Card 组件..................... 97
　8.3.3　Input 组件.................... 99
　8.3.4　List 组件..................... 101
　8.3.5　Search 组件................ 105
　8.3.6　Select 组件................. 110
　8.3.7　TabBottom 组件.......... 115
　8.3.8　Title 组件................... 118
　8.3.9　Type 组件................... 122
　8.3.10　Address 组件............ 124

8.4 界面分析 135
　8.4.1 页面结构 135
　8.4.2 图片元素 135
　8.4.3 页面色彩 135
　8.4.4 页面边距 136

第 9 章　用户模块开发.....137
9.1 注册页面开发 137
9.2 登录页面开发 144

第 10 章　商品模块开发...148
10.1 发布/修改商品页面开发 148
10.2 商品列表/首页开发 153
10.3 商品详情页面开发 162

第 11 章　支付模块开发...171
11.1 订单页面开发 171
11.2 订单详情页面开发 172

第 12 章　消息模块开发...178
12.1 消息列表页面开发 178
12.2 消息详情页面开发 182

第 13 章　个人中心模块
　　　　　开发 192
13.1 个人中心页面开发 192

13.2 编辑个人信息页面开发 196
13.3 我发布的商品列表页面
　　 开发 200
13.4 我发布的商品信息编辑页面
　　 开发 205
13.5 我卖出的商品列表页面
　　 开发 211
13.6 我买到的商品列表页面
　　 开发 214
13.7 修改密码页面开发 218
13.8 忘记密码页面开发 221
13.9 修改手机号页面开发 225
13.10 我的收货地址列表页面
　　　开发 232
13.11 添加/编辑收货地址页面
　　　开发 235
　　13.11.1 添加收货地址 235
　　13.11.2 编辑收货地址 236
　　13.11.3 编写代码 237

第 14 章　前端环境部署...243
项目部署流程 243

第 15 章　前端开发总结...247
15.1 开发思路总结 247
15.2 项目难点总结 247

后端开发

第 16 章　Java 基础..........254
16.1　Java 主要特点 254
16.2　Java 语法......................255
　　16.2.1　数据类型.................. 255
　　16.2.2　标识符..................... 256
　　16.2.3　修饰符..................... 256
　　16.2.4　变量......................... 257
　　16.2.5　运算符..................... 257
　　16.2.6　关键字..................... 260
　　16.2.7　注释......................... 261
16.3　Java 开发环境................262
　　16.3.1　JDK.......................... 262
　　16.3.2　JRE.......................... 262
　　16.3.3　JVM......................... 262
　　16.3.4　配置环境变量.......... 262

第 17 章　Spring Boot 框架....................264
17.1　Spring Boot 概述............264
17.2　Spring Boot 解决的问题........................265
17.3　Spring Boot 核心机制......266
17.4　Spring Boot 优缺点........266

第 18 章　后端工具 / 库 ...268
18.1　Java IDE.........................268
　　18.1.1　常用的 IDE................... 268
　　18.1.2　使用 IntelliJ IDEA 创建 Java 项目...............................269
　　18.1.3　使用 IntelliJ IDEA 创建 Spring Boot 项目.......... 273
18.2　Navicat...........................275
　　18.2.1　版本.......................... 276
　　18.2.2　Navicat for MySQL 使用.............................. 277
18.3　Postman.....................280
　　18.3.1　安装.......................... 281
　　18.3.2　版本.......................... 281
　　18.3.3　使用.......................... 281

第 19 章　后端系统设计....283
19.1　数据库设计....................283
19.2　系统功能模块设计............290
19.3　接口设计......................291

第 20 章　用户模块接口...294
20.1　注册接口........................294
20.2　获取验证码接口................304
20.3　登录接口........................311
20.4　忘记密码接口..................317
20.5　修改密码接口..................321
20.6　用户信息修改接口............324
20.7　校验是否登录接口............328
20.8　退出登录接口..................330

第 21 章　商品模块接口...333

- 21.1　商品类别列表接口...........333
- 21.2　商品列表接口.....................338
- 21.3　发布商品接口.....................346
- 21.4　修改商品信息接口............351
- 21.5　获取商品详情接口............356
- 21.6　评论/回复接口...................361
- 21.7　评论/回复列表接口.........365
- 21.8　点赞/取消点赞接口.........368
- 21.9　点赞列表接口.....................372
- 21.10　首页轮播商品列表接口...375

第 22 章　消息模块接口...382

- 22.1　聊天对话框列表接口.......382
- 22.2　获取聊天详情接口............387
- 22.3　初始化聊天接口.................393
- 22.4　发送消息接口.....................400

第 23 章　支付模块接口...412

- 23.1　购买商品接口.....................412
- 23.2　获取订单详情接口............420
- 23.3　取消订单接口.....................426
- 23.4　支付宝 WAP 支付接口......429
- 23.5　支付宝支付接口.................436
- 23.6　支付宝支付回调接口........440

第 24 章　个人中心模块接口...446

- 24.1　我的商品列表接口............446
- 24.2　删除我的商品接口............455
- 24.3　校验旧手机号接口............459
- 24.4　绑定新手机号接口............462
- 24.5　查询商品数量接口............465
- 24.6　收货地址列表接口............471
- 24.7　新增收货地址接口............478
- 24.8　修改收货地址接口............482
- 24.9　删除收货地址接口............487

第 25 章　后端环境部署...491

第 26 章　后端开发总结...495

- 26.1　开发思路总结.....................495
- 26.2　开发难点总结.....................495

项目概述

第1章 项目需求设计

【本章导读】
◎ 项目背景
◎ 项目介绍

1.1 项目背景

随着互联网行业日趋迅猛的发展，电子商务产业水涨船高。网上交易的便利使得人们购物欲望越来越"膨胀"，购物行为也越来越随意。很多人往往买了一大堆物品，但真正合适或者喜欢的却并不多，那么买到的用不上或者不合适的物品应该怎么处理呢？很多人第一反应是退货或者换货，但是也有一部分人嫌退换货手续麻烦，还要承担运费等损失，于是便选择放置家中，长此以往，堆积的闲置物品越来越多。于是，闲置物品交易平台也顺应市场而生。

1.2 项目介绍

本项目名称为"二手市场"，是一个闲置物品交易平台。在该平台，用户可在"首页"浏览所有人发布的闲置物品信息。如对某一商品感兴趣，可查看商品详情。本项目提供了注册、登录入口，用户可免费注册，登录成功之后，可对商品进行点赞或者留言，也可以与其他用户即时聊天。用户还可通过"发布"功能发布自己的闲置物品信息。项目还包含个人中心模块，通过该模块用户可对个人信息进行编辑，也可查看自己发布的、卖出的、买入的商品信息，还可修改密码或者找回密码。

第 2 章　业务模块介绍

【本章导读】
- ◎ 用户模块
- ◎ 商品模块
- ◎ 支付模块
- ◎ 消息模块
- ◎ 个人中心模块

本项目所有的功能分为五大业务模块，分别为用户模块、商品模块、支付模块、消息模块以及个人中心模块。

用户模块：主要提供用户注册、用户登录、获取验证码、忘记密码等功能。

商品模块：主要提供商品发布、商品信息分页查询、商品关键字查询、商品类别查询、商品点赞及留言等功能。

支付模块：主要提供商品交易支付功能。

消息模块：主要提供查看消息列表、与商品发布者/商品购买者即时聊天等功能。

个人中心模块：主要提供个人信息管理和商品管理等功能。个人信息管理包含个人资料编辑、密码修改等；商品管理包含个人发布的商品列表及详情、个人卖出的商品列表及订单详情、个人买到的商品列表及订单详情等。

2.1　用户模块

用户模块主要提供与用户相关的功能，比如用户注册、用户登录、修改密码、忘记密码等。

2.1.1 用户模块功能清单

用户模块功能清单如表 2-1 所示。

表 2-1 用户模块功能清单

业务模块编号	业务模块名称	功能编号	功能名称	所属页面	子功能编号	子功能描述
M1	用户	M1-01	用户注册	注册页面	M1-01-01	参数非空及正则校验
					M1-01-02	头像上传
					M1-01-03	获取手机验证码
					M1-01-04	密码与确认密码对比校验
					M1-01-05	点击【注册】,若校验通过,静默登录,进入商品发布页面
					M1-01-06	点击【已有账号？登录】,进入登录页面
		M1-02	用户登录	登录页面	M1-02-01	参数非空及正则校验
					M1-02-02	点击【登录】,若校验通过,进入首页,同时本地缓存账号信息
					M1-02-03	点击【快速注册】,进入注册页面
					M1-02-04	点击【忘记密码】,进入忘记密码页面

2.1.2 用户模块功能说明

用户模块包含的功能为用户注册和用户登录。

用户注册包含的子功能说明如下。

（1）头像、昵称、手机号、验证码、密码及确认密码非空校验。

（2）头像上传。

（3）手机号校验规则：第一位以 1 开头，第二位以 3、4、5、7、8 开头，一共由 11 位数字组成。

（4）调取验证码接口才获取手机验证码。

（5）密码与确认密码校验规则：需由数字、字母组成，长度为 6～16，密码与确认密码必须一致。

（6）点击【注册】，若校验通过，静默登录，进入商品发布页面。

（7）点击【已有账号？登录】，进入登录页面。

用户登录包含的子功能说明如下。

（1）账号（昵称或者手机号）、密码非空校验。

（2）密码校验规则：需由数字、字母组成，长度为 6 ~ 16。

（3）点击【登录】，若校验通过，进入首页，同时在本地缓存账号信息。

（4）点击【快速注册】，进入注册页面。

（5）点击【忘记密码】，进入忘记密码页面。

2.2 商品模块

商品模块主要提供与商品相关的功能，比如商品发布、商品查询、商品详情等。

2.2.1 商品模块功能清单

商品模块功能清单如表 2-2 所示。

表 2-2 商品模块功能清单

业务模块编号	业务模块名称	功能编号	功能名称	所属页面	子功能编号	子功能描述
M2	商品	M2-01	商品发布	商品发布页面	M2-01-01	参数非空及正则校验
					M2-01-02	商品图片上传
					M2-01-03	密码与确认密码对比校验
					M2-01-04	商品分类选项弹框
					M2-01-05	点击【发布】，若校验通过，进入首页
		M2-02	商品查询	首页	M2-02-01	获取前 10 条商品信息
					M2-02-02	通过关键字对商品进行查询
					M2-02-03	点击发布图标，进入商品发布页面
					M2-02-04	轮播图
					M2-02-05	通过商品列表对商品进行查询
					M2-02-06	商品信息分页查询
					M2-02-07	点击商品图片，进入商品详情页面
		M2-03	商品详情	商品详情页面	M2-03-01	获取用户头像、昵称、商品发布时间

续表

业务模块编号	业务模块名称	功能编号	功能名称	所属页面	子功能编号	子功能描述
M2	商品	M2-03	商品详情	详情页面	M2-03-02	获取商品信息
					M2-03-03	对商品点赞
					M2-03-04	对商品留言
					M2-03-05	与商品发布人在线聊天

2.2.2 商品模块功能说明

商品模块包含的功能为商品发布、商品查询以及商品详情。

商品发布包含的子功能说明如下。

（1）商品描述、图片、价格、分类非空校验。

（2）商品描述字数限制在1000字（含）以内。

（3）图片数量限制在9张（含）以内。

（4）价格由8位整数、一个小数点、2位小数组成。

（5）商品分类选项弹框。

（6）点击【确认发布】，若商品校验通过，则进入商品列表页面，即首页。

商品查询包含的子功能说明如下。

（1）调取商品查询接口，获取前10条商品信息。

（2）在搜索框输入关键字，可对商品进行关键字查询。

（3）点击发布图标，进入商品发布页面。

（4）实现轮播图。

（5）通过商品列表，对商品进行查询。

（6）商品信息分页查询，每次上拉加载10条产品信息。

（7）点击商品图片，可进入商品详情页，并将商品信息携带过去。

商品详情包含的子功能说明如下。

（1）从本地缓存中获取用户头像、昵称。

（2）从首页的数据中获取商品发布时间、商品描述、商品图片。

（3）看到中意商品，可对商品进行点赞。

（4）看到中意商品或有疑问商品，可对商品留言。

（5）可与商品发布人即时聊天。

2.3 支付模块

支付模块主要提供下单与支付、获取订单详情功能。

2.3.1 支付模块功能清单

支付模块功能清单如表 2-3 所示。

表 2-3 支付模块功能清单

业务模块编号	业务模块名称	功能编号	功能名称	所属页面	子功能编号	子功能描述
M3	支付	M3-01	下单与支付	订单确认页面	M3-01-01	添加收货地址（无任何收货地址时）
					M3-01-02	选择收货地址（有收货地址时）
					M3-01-03	点击【确定】，提交订单
		M3-02	获取订单详情	订单详情页面	M3-02-01	获取订单信息
					M3-02-02	点击【关闭交易】，可关闭当前交易
					M3-02-03	点击【我要付款】，可付款

2.3.2 支付模块功能说明

支付模块包含的功能为下单与支付、获取订单详情。

下单与支付包含的子功能说明如下。

（1）显示下单的商品信息。

（2）如果没有添加过收货地址，在该页面需要添加收货地址信息；如果添加过收货地址，在该页面也可以修改收货地址信息。

（3）订单信息确认完之后，点击【确定】，调取下单接口，在下单的回调里调取支付接口，使用支付宝的支付功能。

获取订单详情包含的子功能说明如下。

（1）获取订单信息，包括价格、图片、描述、收货地址等。

（2）当付款失败或者未付款时，点击【关闭交易】，可关闭当前交易。

（3）当付款失败或者未付款时，点击【立即付款】，可付款。

2.4 消息模块

消息模块主要是方便用户在线咨询、在线聊天，假如用户看到心仪商品或者对商品价格有异议时，可在线与商品发布者即时沟通。

2.4.1 消息模块功能清单

消息模块功能清单如表 2-4 所示。

表 2-4 消息模块功能清单

业务模块编号	业务模块名称	功能编号	功能名称	所属页面	子功能编号	子功能描述
M4	消息	M4-01	消息列表	消息列表页面	M2-01-01	获取消息列表数据
					M2-01-02	点击某一行消息，进入消息详情页面，可在线即时聊天
		M4-02	在线聊天	消息详情页面	M2-02-01	在线即时聊天

2.4.2 消息模块功能说明

消息模块包含的功能为消息列表、在线聊天。

消息列表包含的子功能说明如下。

（1）获取消息列表数据。

（2）点击某一行消息，进入消息详情页面，可在线即时聊天。

在线聊天包含的子功能说明如下：

在线即时聊天。

2.5 个人中心模块

个人中心模块包含个人资料编辑、密码修改、个人发布的商品信息修改；也包含个人发布的商品列表及详情、个人卖出的商品列表及详情以及个人买到的商品列表及详情。即个人中心模块既包含用户信息，也包含与用户相关的商品信息。

2.5.1 个人中心模块功能清单

个人中心模块功能清单如表 2-5 所示。

表 2-5 个人中心模块功能清单

业务模块编号	业务模块名称	功能编号	功能名称	所属页面	子功能编号	子功能描述
M5	个人中心	M5-01	个人中心展示	个人中心页面	M5-01-01	点击用户头像，可进入个人资料编辑页
					M5-01-02	点击【我发布的】，可进入我发布的商品列表页面
					M5-01-03	点击【我卖出的】，可进入我卖出的商品列表页面
					M5-01-04	点击【我买到的】，可进入我买到的商品列表页面
					M5-01-05	点击【修改密码】，可进入修改密码页面
					M5-01-06	点击【退出登录】，返回登录页面
		M5-02	个人资料编辑	编辑个人信息页面	M5-02-01	参数非空及正则校验
					M5-02-02	点击【确认修改】，若校验通过，进入个人中心页面
		M5-03	我发布的	我发布的商品列表页面	M5-03-01	点击某一行，进入我发布的商品详情页面
					M5-03-02	点击【修改】，进入商品发布页面，可对商品信息进行修改
				我发布的商品详情页面	M5-03-03	获取我发布的商品详情信息
		M5-04	我卖出的	我卖出的商品列表页面	M5-04-01	点击某一行，进入我卖出的订单详情页面
					M5-04-02	点击【联系买家】，可与买家即时聊天
				我卖出的订单详情页面	M5-04-03	获取我卖出的订单详情信息
					M5-04-04	点击买家昵称可与买家即时聊天
		M5-05	我买到的	我买到的商品列表页面	M5-05-01	点击某一行，进入我买到的订单详情页面
					M5-05-02	点击【联系卖家】，可与卖家即时聊天
				我买到的订单详情页面	M5-05-03	获取我卖出的订单详情信息
					M5-05-04	点击买家昵称可与买家即时聊天

续表

业务模块编号	业务模块名称	功能编号	功能名称	所属页面	子功能编号	子功能描述
M5	个人中心	M5-06	修改密码	修改密码页面	M5-06-01	参数非空及正则校验
					M5-06-02	点击【确认修改】，若校验通过，进入个人中心页面

2.5.2 个人中心模块功能说明

个人中心模块包含的功能为个人中心展示、个人资料编辑、我发布的、我卖出的、我买到的以及修改密码。

个人中心展示包含的子功能说明如下。

（1）点击用户头像，可进入编辑个人信息页面。

（2）点击【我发布的】，可进入我发布的商品列表页面。

（3）点击【我卖出的】，可进入我卖出的商品列表页面。

（4）点击【我买到的】，可进入我买到的商品列表页面。

（5）点击【修改密码】，可进入修改密码页面。

（6）点击【退出登录】，清除本地缓存，调退出登录接口，成功之后返回登录页面。

个人资料编辑包含的子功能说明如下。

（1）头像、昵称、手机号非空及正则校验。

（2）昵称由6～16位数字和字符组成。

（3）点击【确认修改】，若校验通过，进入个人中心页面。

我发布的包含的子功能说明如下。

（1）在我发布的商品列表页面，点击某一行可进入我发布的商品详情页面。

（2）在我发布的商品列表页面，点击【修改】，进入商品发布页面，可对商品信息进行修改。

（3）在我发布的详情页面，获取从我发布的列表页传递过来的对象集合。

我卖出的包含的子功能说明如下。

（1）在我卖出的商品列表页面，点击某一行可进入我卖出的订单详情页面。

（2）在我卖出的商品列表页面，点击【联系买家】，可与买家即时聊天。

（3）在我卖出的订单详情页面，获取从我卖出的商品列表页面传递过来的对象集合。

（4）在我卖出的订单详情页面，点击买家昵称可与买家即时聊天。

我买到的包含的子功能说明如下。

（1）在我买到的商品列表页面，点击某一行可进入我所买到的订单详情页。

（2）在我买到的商品列表页面，点击【联系卖家】，可与卖家即时聊天。

（3）在我买到的订单详情页面，获取从我买到的商品列表页面传递过来的对象集合。

（4）在我买到的订单详情页面，点击卖家昵称可与卖家即时聊天。

修改密码包含的子功能说明如下。

（1）手机号、密码及确认密码非空及正则校验。

（2）手机号校验规则、密码及确认密码校验规则与前面保持一致。

（3）点击【确认修改】，若校验通过，进入个人中心页面。

第 3 章　业务流程图

【本章导读】
- ◎ 注册页面流程图
- ◎ 登录页面流程图
- ◎ 商品发布页面流程图
- ◎ 首页流程图
- ◎ 消息页面流程图
- ◎ 个人中心页面流程图

3.1　注册页面流程图

图 3-1 所示为注册页流程图。在登录页面，用户可点击快速【注册】按钮进入注册

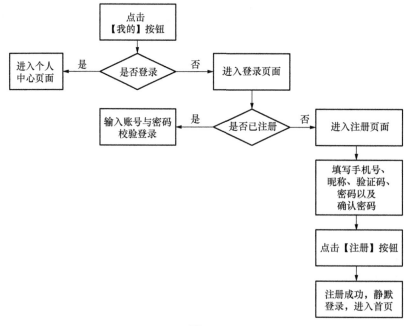

图 3-1

页面；在注册页面，需要填写手机号（并发送验证码以验证有效手机号）、昵称、验证码、密码、确认密码进行注册。注册成功之后静默登录（用户状态更新为登录状态），进入首页。

3.2 登录页面流程图

图 3-2 所示为登录页流程图。用户点击【我的】按钮，检测是否是登录状态，如果未登录，则进入登录页面，可输入账号和密码进行登录。

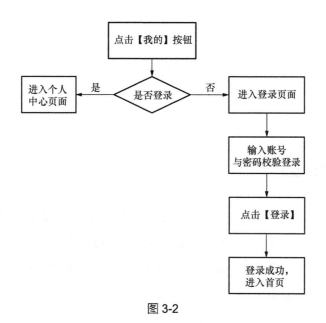

图 3-2

3.3 商品发布页面流程图

图 3-3 所示为发布页流程图。用户可通过点击【发布】按钮，进入商品发布页面。在商品发布页面，也会检测用户登录状态。登录用户可以在商品发布页面上传商品图片，填写商品描述、商品价格、商品分类等信息。填写完成之后，可点击【发布】按钮将商品上架，随后跳转到首页。

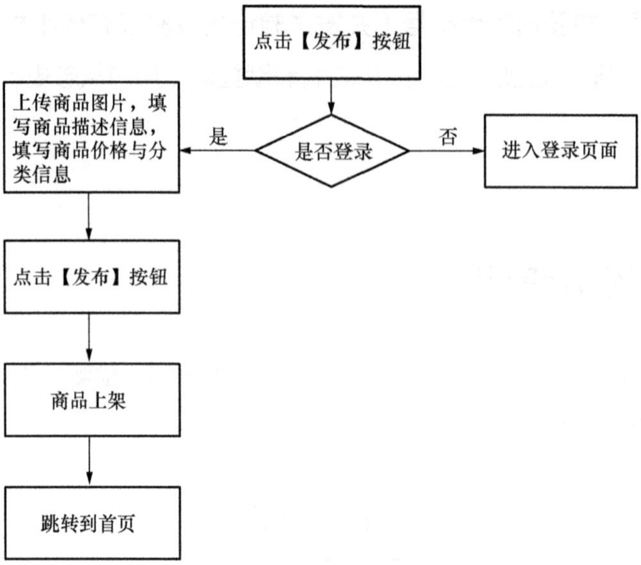

图 3-3

3.4 首页流程图

图 3-4 所示为首页流程图。打开项目后会默认展示首页，在首页可浏览商品列表信息，可点击感兴趣的某个商品，进入商品详情页。在商品详情页，如果是已经登录的用户，可对商品进行点赞或者留言，也可以通过聊一聊功能与商品发布者即时在线沟通；如果未登录，则会先进入登录页面。

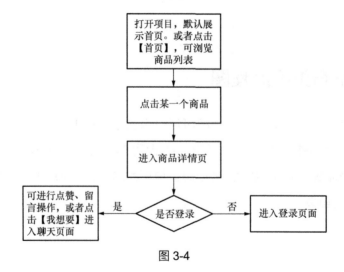

图 3-4

3.5 消息页面流程图

图 3-5 所示为消息页面流程图。用户点击【消息】按钮，检测是否登录，如果已登录，则可查看消息列表，点击某一条消息，进入聊天页面，可发送和接收消息。如果未登录，则会先进入登录页面。

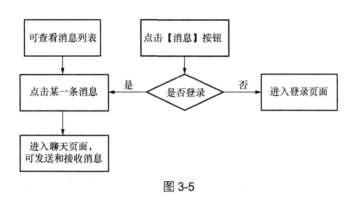

图 3-5

3.6 个人中心页面流程图

图 3-6 所示为个人中心页面流程图。用户点击【我的】按钮，如果是登录状态，则跳转到个人中心页面。在个人中心页面，分别有 8 个功能点。

（1）点击头像，进入编辑个人信息页面，可对昵称和手机号进行修改，修改完之后，点击确认修改，进入个人中心页面。

（2）点击【我发布的】，进入我发布的商品列表页面，点击某一条记录，进入我发布的商品详情页面。

（3）点击【我卖出的】，进入我卖出的商品列表页面，点击某一条记录，进入我卖出的订单详情页面。

（4）点击【我买到的】，进入我买到的商品列表页面，点击某一条记录，进入我买到的订单详情页面。

（5）点击【修改密码】，进入修改密码页面，分别输入原密码、新密码与确认密码，点击确认修改，进入个人中心页面。

（6）点击【忘记密码】，进入忘记密码页面，分别输入手机号、验证码、新密码与确认密码，点击完成，进入个人中心页面。

（7）点击【在线客服】，弹出直接拨打客服电话的界面。

（8）点击【退出登录】，进入登录页面。

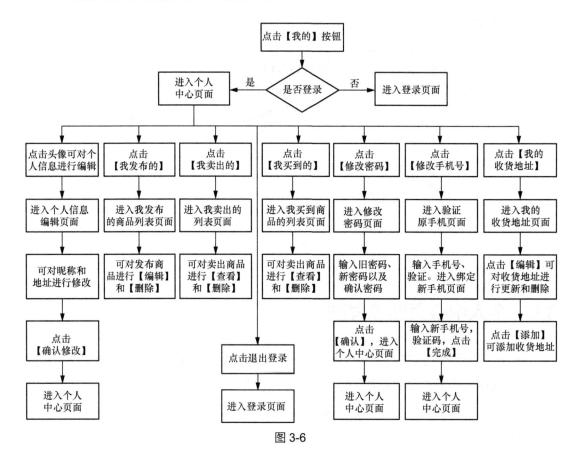

图 3-6

第 4 章　项目开发周期

【本章导读】

◎ 项目开发周期

项目开发周期

在统筹一个完整项目时，会分不同阶段来规划整个项目。一般情况下分为需求确认阶段、项目开发阶段和项目部署阶段。这些都需要项目负责人统一安排协调。

在需求确认阶段，主要确定项目需求、项目方案以及技术选型等。在正式投入开发前会有项目需求探讨阶段。由项目负责人、UI 设计师、前后端开发工程师一起参与讨论。项目需求确定之后，再由项目负责人拟定项目实施方案。由开发者确定技术选型。

在项目开发阶段，该阶段需求基本确认完成。首先由 UI 设计师将各功能模块的页面设计完成，然后由前端开发工程师根据 UI 设计进行静态页面开发。后端开发工程师也可以根据 UI 设计和业务逻辑进行数据库设计。前面说到，本项目可以由一个人进行开发，读者可能会疑惑，作为开发人员不懂 UI 设计怎么办？其实现在市面上有很多原型设计工具，非常好用，不会设计的开发者可以使用这些工具，很快就能设计出基础的原型图。本文使用的是 Axure 原型设计工具。

在项目部署阶段，项目部署也会分为 3 个阶段。第一阶段是开发部署。该阶段是项目基本开发完成，可由开发人员自行部署测试。第二阶段是测试部署。测试部署环境分为 SIT（System Integration Testing，系统集成测试）环境和 UAT（User Acceptance Testing，用户验收测试）环境。该阶段一般由项目负责人部署。第三阶段是上线部署。一般由部门最高负责人或者运维人员进行部署。代码一经投产，便不可再更改。

以下是按一个人全天设计开发本项目的时间规划，仅供参考。

表 4-1　Axure 原型设计时间规划表

模块	设计时间	页面
账户模块	1 周	注册页面
		登录页面
产品模块		商品发布页面
		商品列表（首）页
		商品详情页
支付模块		支付页面
		订单页面
消息模块		消息列表页面
		消息详情页面
个人中心模块	1 周	个人中心页面
		个人信息编辑页面
		我发布的列表页面
		我发布的详情页面
		我卖出的列表页面
		我卖出的详情页面
		我买入的列表页面
		我买入的详情页面
		修改密码页面
		修改手机号页面
		编辑收货地址页面

表 4-2　前端开发时间规划表

模块	开发时间	页面
账户模块	1 周	注册页面
		登录页面
产品模块	1 周	商品发布页面
		商品列表页面
		商品详情页面
支付模块	1 周	支付页面
		订单页面
消息模块	1 周	消息列表页面
		消息详情页面

续表

模块	开发时间	页面
个人中心模块	2周	个人中心页面
		个人信息编辑页面
		我发布的列表页面
		我发布的详情页面
		我卖出的列表页面
		我卖出的详情页面
		我买入的列表页面
		我买入的详情页面
		修改密码页面
		修改手机号页面
		编辑收货地址页面

表 4-3 后端开发时间规划表

模块	开发时间	接口
用户模块	1周	获取验证码接口
		注册接口
		登录接口
商品模块	1周	获取点赞列表接口
		点赞接口
		获取留言列表接口
		留言接口
		商品发布接口
		获取商品列表接口
		获取商品类型接口
		获取商品详情接口
		获取点赞列表接口
		获取留言列表接口
		点赞接口
		留言接口
		商品发布接口
订单模块	1周	下单接口
		支付接口

续表

模块	开发时间	接口
消息模块	1周	获取消息列表接口
		发送消息接口
		获取消息接口
个人中心模块	2周	个人商品列表查询
		发布商品修改接口
		发布商品删除接口
		修改密码接口
		修改手机号接口
		添加、删除、编辑接口

前端开发

第 5 章 HTML、CSS、JavaScript 基础

【本章导读】
◎ HTML 历史、HTML 结构、HTML 常用标签
◎ CSS 历史、CSS 样式、CSS 选择器、CSS 伪类与伪元素、CSS 使用方式
◎ JavaScript 历史、JavaScript 语法以及 ATAX
◎ 编写 Web 网页案例

说到前端开发，自然离不开"网页三剑客"——HTML、CSS、JavaScript。目前市面上实际的前端开发都是以框架驱动，但是框架的底层实现还是离不开基础。作为一名合格的前端开发者，不仅要会熟练使用框架，更要牢牢掌握网页开发基础。本章将会介绍 HTML、CSS、JavaScript 3 门语言的前世今生，诞生缘由，还会列出一些常见的 HTML 标签和 CSS 样式以及 JavaScript 语法。最后会编写一个 Web 网页案例使读者能够学以致用、融会贯通。

5.1 网页的骨架——HTML

HTML 全称为 HyperText Markup Language，中文译为超文本标记语言，"超文本"指的就是并非单纯的文字，也包含一些图片、超链接、视频、音频等。它并非编程语言，而是标记语言。其结构主要分为头部和主体部分。头部主要用来描述网页信息，比如网页标题和资源引用。主体部分就是网页的具体内容。HTML 的核心元素是标签，标签是组成页面元素的最小单元。页面元素通常是由开始标签、具体内容、结束标签组成，比如 <div> 我是有结束标签的元素 </div>。也有一些没有结束标签的标签元素，比如 <input value="我是没有结束标签的元素 "/>。结束标签也叫"闭合标签"。

5.1.1 HTML 结构

一个完整的 HTML 页面由 <!DOCTYPE>（DOCUMENT TYPE，即文档类型）和最顶层的父标签 <html></html>（父标签内有头部标签、样式标签、内容标签及脚本标签）组成。

```
<!DOCTYPE html>
<html lang="en">
<head>
    <meta charset="utf-8"/>
    <title></title>
</head>
<style></style>
<body></body>
<script></script>
</html>
```

如上所示，这是一个最简单的空 HTML 网页。其中 <!DOCTYPE html> 为 HTML5 的文档类型声明；html 标签为父标签，包裹所有的标签内容；head 标签为头部标签，用来描述网页信息；style 标签为样式标签，用来定义网页样式；body 标签为内容标签，用来定义网页内容；script 标签为脚本标签，用来定义网页脚本，实现网页交互功能。

1. <!DOCTYPE>

<!DOCTYPE> 的作用是声明应该用什么样的文档类型定义来解析网页。在 HTML 4.01 中，<!DOCTYPE> 声明引用 DTD（Document Type Definition，文档类型定义），因为 HTML 4.01 是基于 SGML 开发的。而 DTD 规定了标记语言的规则，这样浏览器才能根据声明的类型来解析响应的文档内容。

在 HTML 4.01 中有 3 种 <!DOCTYPE> 声明。

第一种是严格模式，包含所有的 HTML 元素和属性，但不包括展示性的和弃用的元素，也不包含框架集。

如 <!DOCTYPE HTML PUBLIC "-//W3C//DTD HTML 4.01//EN" "http : //www.w3.org/TR/html4/strict.dtd">。

第二种是传统模式，包含所有的 HTML 元素和属性，既包括展示性的元素，也包括弃用的元素，但不包含框架集。

如 <!DOCTYPE HTML PUBLIC "-//W3C//DTD HTML 4.01 Transitional//EN" "http://www.w3.org/TR/html4/loose.dtd">。

第三种是框架模式，既包含所有的 HTML 元素和属性，也包含框架集。

如 <!DOCTYPE HTML PUBLIC "-//W3C//DTD HTML 4.01 Frameset//EN" "http://www.w3.org/TR/html4/frameset.dtd">。

因为 HTML5 不基于 SGML，所以不需要引用 DTD。在 HTML5 中只需要写成 <!DOCTYPE html> 即可。

2. <html></html>

html 标签是父标签，由 head 标签、style 标签、body 标签及 script 标签组成。

5.1.2　HTML 头部标签

head 标签用来定义文档的头部信息，head 标签作为父标签，其子标签常见的有 meta 标签、title 标签、link 标签等。

1. meta 标签

meta 标签用来定义 HTML 文档中的元数据，其中 3 个属性分别为 charset、name、http-equiv。

charset：用来定义文档采用的编码方式。常用的是 UTF-8。

name：用来描述网页的内容信息，如 author、keywords、description 等。name 通常与 content 结合使用，name 中的值为键名，content 里面的值为键值。比如开发移动端应用时定义视口。手机浏览器把页面放在一个虚拟的"窗口"（viewport）中，这个窗口就是视口。设置 name="viewport" 可以让网页开发者控制视口的大小和缩放比例。此时 content 中可以设置如下 6 个属性。

　　　　width：指定宽度，不带单位，值为 device-width 时表示与屏幕同宽。

　　　　height：指定高度，和 width 相对应。

　　　　initial-scale：初始缩放比例，即页面第一次加载的时候缩放比例。值为 1.0 时，表示不缩放。

　　　　maximum-scale：允许用户缩放到的最大比例。

　　　　minimum-scale：允许用户缩放到的最小比例。

　　　　user-scalable：布尔值，取值为 yes 或者 no，表示用户是否可以手动缩放。

如 <meta name="viewport" content="width=device-width, initial-scale=1.0">。

http-equiv：用来向浏览器传输有关网页的信息，并精确地显示内容。同样与 content 结合使用。http-equiv 取值有 expires（设置网页过期时间）、Pragma（是否禁止浏览器从本地缓存中访问内容）、Refresh（自动刷新并指向新页面）、Set-Cookie（如果网页过期，那么存储的 cookies 将被删除）、Window-target（显示窗口的设定）、content-Type（显示字符集的设定）、content-Language（显示语言的设定）、Cache-Control（指定请求和响应遵循的缓存机制）。

如：

<meta charset="utf-8"/>，采用 UTF-8 的编码格式，可以忽略大小写；

<meta name="viewport" content="width=device-width, initial-scale=1, user-scalable=no" />，与屏幕同宽，缩放比例为 1，不允许用户手动缩放；

<meta name="theme-color"content="#000000" />，采用颜色值为 #000000 的颜色作为主题颜色。

2. title 标签

title 标签很容易理解，就是定义文档的标题。比如打开一个网页，在最顶部网页切换的位置看到的内容就是 title 标签所定义的内容。

如 <title> 我是网页标题 </title>。

3. link 标签

link 标签用来定义文档和外部资源之间的关系。通常用来引用外部 CSS 样式表。

如 <link rel="stylesheet" href="style.css">。

5.1.3　HTML 内容标签

HTML 内容标签即 <body></body> 标签，该标签中承载了文档的主体内容。body 标签是内容标签的父级标签，其子标签有很多种。

（1）按照是否闭合分为非闭合标签和闭合标签。

非闭合标签内不需要嵌入文本，也叫无内容标签，通常有两种写法，一种是
，另一种是
，前者是 HTML 的写法，后者是 XHTML 的写法。常见非闭合标签如下。

<input type="text">：文本输入框标签。

<img src=""alt="">：图片标签。

：换行标签。

<hr>：水平分割线标签。

<link>：引入外部资源标签。

<meta>：定义文档元信息标签。

除了非闭合标签，其他都是闭合标签。

（2）按照是否换行显示分为块级标签和行内标签。

块级标签特征：宽度占满屏幕，垂直显示，可以设置高宽值，块级标签大多为结构性标记（定义元素结构）。

块级标签如下。

<div></div>：最常用的块级标签，无任何语义，可定义文档中的分区或节。

<p></p>：定义段落的标签。

<blockquote></blockquote>：定义段落缩进的标签。

<marquee behavior=""direction=""></marquee>：定义滚动文本的标签。

<address> </address>：定义地址标签，默认为斜体。

<ul></ul>：定义无序列表标签。

<ol></ol>：定义有序列表标签。

<li></li>：定义列表项目标签。

<dl></dl>：定义列表标签。

<dt></dt>：定义列表中的描述标签。

<dd></dd>：定义列表中的条目标签。

<table></table>：定义表格标签。

<form action=""></form>：定义表单标签。

<hr>：定义水平分割线标签。

<h1> ~ <h6>定义标题标签。

行内标签特征：行内标签与块级标签正好相反，不会占满整个屏幕，因此行内标签会水平显示，不可设置高、宽值，行内标签大多为描述性标记（定义元素样式）。

行内标签如下。

<i></i>：定义斜体标签。

<em></em>：定义强调文本，斜体加黑体标签。

<del></del>：定义删除线标签。

<sub></sub>：定义上标标签。

<sup></sup>：定义下标标签。

<b></b>：定义加粗标签（属于 HTML 标签）。

<strong></strong>：定义强调文本标签，默认文本会加粗（属于 XHTML 标签）

<span></span>：无任何语义，用来组合文档中的行内元素。

：定义换行标签。

<label for=""></label>：为 nput 元素定义标注（标记）。

<input type="text">：定义单行文本输入框标签，适合数值及短文本。

<textarea name="" id="" cols="30" rows="10"></textarea>：定义多行文本输入框标签，适合长文本。

<a href=""></a>：定义超链接标签。

<select name="" id=""></select>：定义下拉列表标签。

<img src="" alt="">：定义图片标签。

注意1：以上列举的标签只是常见的一些标签，并非所有标签，而且随着 W3C 的不断更新改进，会新增和废弃部分标签。

注意2：有几个比较特殊的行内标签可以设置宽和高，分别为 textarea、input、img、select。

注意3：块级标签与行内标签可以相互转换。块级标签设置 {display：inline-block} 可转换为行内标签，具备行业标签的特征，比如可以水平显示。行内标签设置 {display：block} 可转换为块级标签，具备块级标签的特征，比如可以设置高和宽。

5.1.4　HTML 样式标签

HTML 文档中的样式标签是 <style></style>，一般置于 head 标签之后，body 标签之前。我们可在该标签内根据元素选择器定义元素样式。

如下案例所示，style 标签内部定义了 div 元素选择器，将其元素颜色设置为红色，因此文档将显示一段红色的文本。关于文档样式的内容将在 5.2 节中详细介绍。

```
<!DOCTYPE html>
<html lang="en">
<head>
    <meta charset="utf-8"/>
```

```html
        <title></title>
        <link rel="stylesheet" href="">
</head>
<style>
    div{color:red}
</style>
<body>
<div> 我是一段红色文本 </div>
</body>
</html>
```

5.1.5　HTML 脚本标签

HTML 文档中的脚本标签是 <script></script>，一般置于 body 标签之后。我们可在该标签内编写网页脚本，也可以通过该标签的 src 属性引入外部脚本资源。

script 标签的属性有 type（描述脚本的内容类型）、src（引入外部脚本文件的 URL）、async（是否异步执行脚本）、defer（是否对脚本进行延迟，直到页面加载为止）、charset（规定在外部脚本中使用的字符编码）等。

如下案例所示，通过 script 标签的 src 属性引入了 jQuery。引入了 jQuery 之后，就可使用 $ 变量符是 id 是 str 是 div 元素赋值。关于 JavaScript 的具体用法将在 5.3 节中详细介绍。

```html
<!DOCTYPE html>
<html lang="en">
<head>
    <meta charset="utf-8"/>
    <title></title>
    <link rel="stylesheet" href="">
</head>
<style>
    div{
        color:red;
    }
</style>
<style></style>
<body>
<div id="str"></div>
</body>
```

```
<script src="jquery.min.js"></script>
<script>
var content = " 我是一段红色文本 ";
$("#str").html(content);
</script>
</html>
```

5.2 网页的外衣——CSS

CSS 全称为 Cascading Style Sheets，中文译为层叠样式表，用来对 HTML 和 XML 标记语言做格式化处理。比如在网页上呈现出合理的布局，适宜的色彩，还有一些简单的动画效果，都得益于 CSS 才能实现。

5.2.1 CSS 历史

1. CSS 诞生（1994 年）

从 HTML 被创建起，网页样式就以各种各样的形式存在。不同的浏览器有不同的样式语言来控制页面展示效果。起初，HTML 只包含少量的样式标签，后来随着 HTML 的发展，为了满足网页设计者的要求，HTML 提供了很多代表样式的功能标签，因此 HTML 变得越来越"臃肿"。为了使内容与样式能够分离，各自维护，于是 CSS 诞生了。

1994 年，哈肯·维姆·莱（Hakon Wium Lie）在芝加哥的一次会议上第一次提出了 CSS 的建议。当时，伯特·波斯（Bert Bos）正在设计一个叫作 Argo 的浏览器，他们决定一起合作设计 CSS。1995 年，波斯加入 W3C，维姆与波斯一起，再一次提出了这个建议。在 1994 年 W3C 成立时，CSS 的创作成员全部成为了 W3C 的工作小组成员并且全力以赴负责研发 CSS 标准，层叠样式表的开发终于走上正轨。有越来越多的成员参与其中，例如微软公司的托马斯·莱尔顿（Thomas Reaxdon），他的努力最终使 Internet Explorer 支持 CSS 标准。

2. CSS1.0（1996 年）

1996 年 12 月，层叠样式表的第一份正式标准（Cascading Style Sheets Level 1）完成，成为 W3C 的推荐标准。1997 年，W3C 发布 CSS 1.0，CSS 1.0 全面地规定了文档的显示样式，可分为选择器、样式属性、伪类、对象等几个部分。

3. CSS 2.0（1998 年）

1997 年年初，W3C 组织负责 CSS 的工作组开始讨论第一版中没有涉及的问题。其讨论结果组成了 1998 年 5 月发布的 CSS 2.0，目前的主流浏览器都采用这个标准。后来，工作组成员修改了 CSS2.0 中的一些错误，删除了其中不被支持的内容，增加了一些已有的浏览器的扩展内容推出了 CSS2.1。

4. CSS 3.0（2011 年）

从 2011 年开始 CSS 被分为多个模块单独升级，统称为 CSS3。这些模块有 CSS 选择器 level 3、CSS 媒体查询 level 3、CSScolor level 3。这里的 level 3 就是指的 3.0 版本。

5.2.2　CSS 样式

CSS 样式分为基础样式、框模型、布局、定位、动画等。

1. 基础样式

基础样式包括文本或者文本框颜色、背景色、字体、字体大小、是否加粗等。

color：red 表示颜色为红色。

background：red 表示背景色为红色，取值 linear-gradient（red, blue）为渐变色。

font-family：sans-serif 表示字体为 sans-serif。

font-size：14px 表示字体大小为 14px。

font-weight：bold 表示字体加粗，常用的取值有 100 ~ 900、bold（相当于 700）、nomal（相当于 400）。

opacity：0.5 表示透明度为 0.5。

border-radius：50% 表示边框圆角度数为 50%，50% 即圆形。

transform：rotate（30deg）表示样式转换，取值 rotate（30deg）表示旋转 30°。

transition：width 2s 表示样式过渡（元素从一种样式逐渐变成另一种样式的效果）。

2. 框模型

框模型包括元素内容、边框、外边距及内边距。

width：100% 表示宽度填满元素。

height：100% 表示高度填满元素。

border：1px solid red 表示边框宽度为 1px，颜色为红色。

margin：10px 表示外边距，上、下、左、右均为 10px。

margin-left：10px 表示左外边距 10px。

margin-right：10px 表示右外边距 10px。

margin-top：10px 表示上外边距 10px。

margin-bottom：10px 表示下外边距 10px。

padding：10px 表示内边距，上、下、左、右均为 10px。

padding-left：10px 表示左内边距 10px。

padding-right：10px 表示右内边距 10px。

padding-top：10px 表示上内边距 10px。

padding-bottom：10px 表示下内边距 10px。

3. 布局

text-align：left 表示居左显示。

text-align：center 表示居中显示。

text-align：right 表示居右显示。

display：block 使行内元素具备块级元素属性。

display：inline-block 使块级元素具备行内元素属性。

display：flex 表示弹性布局，默认为水平布局。

flex-direction：row/column 表示弹性布局方向，row 为横向，column 为纵向。

flex-wrap：wrap/ nowrap 设置元素超出弹性盒子是否换行。

justify-content：center 设置元素水平方向居中。

justify-content：flex-start 设置元素水平方向居左显示。

justify-content：flex-end 设置元素水平方向居右显示。

justify-content：space-around 设置元素水平方向左右空隙平均分布。

justify-content：space-evenly 设置元素水平方向整体空隙平均分布。

justify-content：space-between 设置元素水平方向靠边分布。

align-items：center 设置元素垂直方向居中。

align-items：flex-start 设置元素垂直方向居上显示。

align-items：flex-end 设置元素垂直方向居下显示。

align-items：baseline 设置元素垂直方向对齐下基准线。

4. 定位

定位指的是元素相对于文档的位置。

定位分别有相对定位（relative）、绝对定位（absolute）、固定定位（fixed）及浮动定位（float）。

position：relative，相对定位是相对于元素本身定位。相对定位无论是否移动，元素仍然会占据原来空间，因此移动元素会导致覆盖其他元素。

position：absolute，绝对定位是相对于最近的已经定位的父元素定位。绝对定位会脱离文档流，因此不会占据空间。

position：fixed，固定定位是相对于文档定位，元素会固定在页面某个位置。固定定位也会脱离文档流，因此不会占据空间。

float：left/right/both，浮动定位不在文档的普通流中，所以文档普通流中的元素会表现出浮动定位元素不存在一样。但是浮动框可以使文本围绕图像。

5. 动画

动画是使元素从一种样式逐渐变化为另一种样式的效果。下面的案例是把 myfirst 动画捆绑到 div 元素，最后的 from 和 to 表示从 0% ~ 100% 的转变。

```
div{ animation:myfirst 5s; }
@keyframes myfirst{
from {background:red;}
to {background:yellow;}
}
```

动画属性如下。

animation-name：myfirst 表示动画名称。

animation-duration：5s 表示动画间隔时间，默认是 0。

animation-timing-function：linear 表示动画的速度曲线，默认是 ease。

animation-delay：2s 表示动画何时开始，默认是 0。

animation-iteration-count：infinite 表示动画播放的次数，infinite 表示无限播放，默认是 1。

animation-direction：alternate 表示动画是否在下一个周期逆向播放，默认是 normal。

animation-play-state：running 表示动画是否正在进行或暂停，默认是 running。

以上属性可以简写为如下形式。

animation：myfirst 5s linear 2s infinite alternate running。

5.2.3　CSS 选择器

CSS 选择器有以下几种。

1. 标签选择器

标签选择器，即使用标签名称来定位元素。标签选择器优先级最低，值为 1。

2. class 类选择器

class 类选择器用" . "表示，class 类选择器优先级值为 10，是使用频率最高的选择器。同一个元素可以有多个 class 类，表现形式为 class=" 类名称 "。

3. id 选择器

id 选择器，用" # "表示。id 选择器是唯一的选择器，同一个文档内容的 id 不可重复。其优先级较高，值为 100。表现形式为 id="id 名称 "。

4. * 选择器

* 代表匹配所有元素。

```
<!DOCTYPE html>
<html lang="en">
<head>
    <meta charset="utf-8"/>
    <title>React</title>
</head>
<style>
    /* 星号选择器，匹配所有元素 */
    * {
        margin:0;
        padding:0;
    }
    /* 标签选择器，匹配所有 div 标签 */
    div {
        font-size:18px;
    }
    /*class 类选择器，匹配 class 名为 item 的标签 */
    .item{
        background:red;
    }
    /*id 选择器，匹配 id 为 only 的标签 */
    #only{
        background:pink;
    }
</style>
<body>
```

```
<div>
    <div class="item">我是 class 类选择器，我不是唯一的，可以有多个重复的 </div>
    <div class="item">我是 class 类选择器，我不是唯一的，可以有多个重复的 </div>
    <div id="only">我是 id 选择器，我是唯一的，只能有一个 </div>
</div>
</body>
</html>
```

5. 属性选择器

① [attribute]：选择带有某个属性的所有元素。

② [attribute = value]：选择属性等于 value 的所有元素。

③ [attribute ~= value]：选择属性中包含 value 元素的所有元素（适用于由空格分隔的属性值）。

④ [attribute |= value]：选择属性中以 value 开头的所有元素，且必须选取整个单词或中间有连字符的单词。

⑤ [attribute ^= value]：选择属性中以 value 开头的所有元素，但不受单词的限制。

⑥ [attribute $= value]：选择属性中以 value 结尾的所有元素。

⑦ [attribute *= value]：选择属性中包含 value 的所有元素。

```
<!DOCTYPE html>
<html lang="en">
<head>
    <meta charset="utf-8"/>
    <title>React</title>
</head>
<style>
    /* 匹配 type 属性的所有元素 */
    [type] {
        color:red;
    }
    /* 匹配 type 属性值为 abc 的元素 */
    [type=abc] {
        color:orange;
    }
    /* 匹配 type 属性值有 mm 的元素（属性值由空格分开）*/
    [type~=mm] {
        color:yellow;
```

```
            }
            /* 匹配 type 属性值以 nn 开头的元素（属性值由"-"分开）*/
            [type|=nn] {
                color:green;
            }
            /* 匹配 type 属性值以 xyz 开头的元素 */
            [type^=xyz] {
                color:cyan;
            }
            /* 匹配 type 属性值以 ef 结尾的元素 */
            [type$=ef] {
                color:blue;
            }
            /* 匹配 type 属性值包含 d 的元素 */
            [type*=d] {
                color:purple;
            }
    </style>
    <body>
    <div>
        <div type="">我有 type 属性，我是红色 </div>
        <div type="abc">我的 type 属性值为 abc，我是橙色 </div>
        <div type="mm ff">我的 type 属性值为 mm ff，我是黄色 </div>
        <div type="nn-ff">我的 type 属性值为 nn-ff，我是绿色 </div>
        <div type="xyzff">我的 type 属性值以 xyz 开头，我是青色 </div>
        <div type="abcgef">我的 type 属性值以 ef 结尾，我是蓝色 </div>
        <div type="abcd">我的 type 属性值包含 d，我是紫色 </div>
    </div>
    </body>
    </html>
```

6. 派生选择器

① 后代选择器：只要是元素的后代元素，都可以包含进来，中间用空格表示。

② 子元素选择器：元素的子元素可以包含进来，用">"表示。

③ 相邻兄弟选择器：如果要选择两个相邻元素，且属于同一个父元素，可以用"相邻兄弟选择器"，相邻兄弟选择器用"+"表示。

```
<!DOCTYPE html>
<html lang="en">
```

```html
<head>
    <meta charset="utf-8"/>
    <title>React</title>
</head>
<style>
    /* 子元素选择器 */
    #container>.item1{
        font-size:20px;
        color:red;
    }
    /* 后代元素选择器 */
    #container span{
        font-size:30px;
        color:blue;
    }
    /* 相邻兄弟元素选择器 */
    .item1+.item2{
        font-size:40px;
        color:yellow;
    }
</style>
<body>
<div id="container">
    <div class="item1">我的父级元素是 id 为 container 的元素，你可以通过子元素选择器找到我
    <span>我的父级元素是 class 为 item 的元素，父级的父级元素是 id 为 ontainer 的元素，你可以通过后代元素选择器找到我 </span>
    </div>
    <div class="item2">我的兄弟元素是 class 为 item 的元素，你可以通过相邻兄弟选择器找到我 </div>
</div>
</body>
</html>
```

5.2.4 CSS 伪类与伪元素

1. CSS 伪类

CSS 伪类用于向某些选择器添加特殊的效果。

CSS 伪类中锚伪类包括 :link（表示未访问的链接）、:visited（表示已访问的链接）、

:hover（鼠标指针悬停效果）、:active（表示选定的链接）等。

在 CSS 定义中，:hover 必须被置于 :link 和 :visited 之后才是有效的。:active 必须置于 : hover 之后才是有效的。伪类不区分大小写。

```
<!DOCTYPE html>
<html lang="en">
<head>
    <meta charset="utf-8"/>
    <title>React</title>
</head>
<style>
/* 未访问的链接 */
    a:link {
color:red
}

/* 已访问的链接 */
    a:visited {color:green}

/* 鼠标指针移动到链接上 */
a:hover {color:purple}

/* 选定的链接 */
a:active {color:blue}
</style>
<body>
<div id="container">
    <a href="http://www.baidu.com">点击此处跳转到百度</a>
</div>
</body>
</html>
```

CSS 伪类中伪类元素选择器包括 :first-child（选择第一个子元素）、:last-child（选择最后一个子元素）、:lang（为不同的语言定义特殊的规则）、:root（选择文档的根元素）、:nth-child（n）（选择某个子元素）等。

```
<!DOCTYPE html>
<html lang="en">
```

```html
<head>
    <meta charset="utf-8"/>
    <title>React</title>
</head>
<style>
/* 选择第一个子元素 */
    #container>div:first-child{
        color:red;
    }
/* 选择某个子元素 */
    #container>div:nth-child(2){
        color:yellow;
    }
/* 选择最后一个子元素 */
    #container>div:last-child{
        color:blue;
    }
</style>
<body>
<div id="container">
    <div>我是第一个子元素，我是红色</div>
    <div>我是第二个子元素，我是黄色</div>
    <div>我是最后一个子元素，我是蓝色</div>
</div>
</body>
</html>
```

2. CSS 伪元素

CSS 伪元素用于向某些选择器设置特殊的效果。

CSS 伪元素包括 :first-line（用于向首行设置不用的效果）、:first-letter（用于向首字母设置不用的效果）、:before（用于在元素的前面插入新内容）、:after（用于在元素的后面插入新内容）等。

```html
<!DOCTYPE html>
<html lang="en">
<head>
    <meta charset="utf-8"/>
    <title>React</title>
```

```
</head>
<style>
    /*设置所有元素颜色为blue*/
    #container{
        color:blue;
    }
    /*将第一行的元素颜色设置为yellow*/
    #container > div:first-line {
        color:yellow;
    }
    /*将第一行第一个字符的元素颜色设置为red,字体大小设置为30px*/
    #container > div:first-letter{
        color:red;
        font-size:30px;
    }
    /* 以下样式的结合设置成文件选择框 */
    #container > #add{
        border:1px solid #ddd;
        position:relative;
        top:50%;
        left:50%;
        width:200px;
        height:200px;
        transform:translate(-50%,50%);
    }
    #container > #add:before{
        background:#ddd;
        content:"";
        width:100px;
        height:5px;
        position:absolute;
        transform:translate(-50%,-50%);
        top:50%;
        left:50%;
    }
    #container > #add:after{
        background:#ddd;
        content:"";
        width:5px;
        height:100px;
        position:absolute;
```

```
            transform:translate(-50%,-50%);
            top:50%;
            left:50%;
        }
    </style>
    <body>
    <div id="container">
        <div>Lorem ipsum dolor sit amet, consectetur adipisicing elit. Atque culpa cum dicta enim facilis fugiat illo iure
            laboriosam, non quaerat quam quidem quis repellendus repudiandae rerum temporibus voluptate voluptates
            voluptatibus?Lorem ipsum dolor sit amet, consectetur adipisicing elit. Assumenda beatae commodi, debitis error
        </div>
        <div id="add"></div>
    </div>
    </body>
    </html>
```

5.2.5　CSS 使用方式

CSS 有 3 种最常见的使用方式。

（1）直接在 HTML 标签中添加样式。

如下案例所示，运行之后会显示一段字体大小为 16px 的文本，字体大小是在元素内联样式中设置。

```
<!DOCTYPE html>
<html lang="en">
<head>
    <meta charset="utf-8"/>
    <title></title>
    <link rel="stylesheet" href="style.css">
</head>
<body>
<div style="font-size:16px">我是一段红色的文本</div>
</body>
</html>
```

（2）在 style 标签中根据选择器属性来添加样式。

如下案例所示，运行之后会显示一段红色的文本，红色是在 style 标签中设置。

```
<!DOCTYPE html>
<html lang="en">
<head>
    <meta charset="utf-8"/>
    <title></title>
    </head>
<body>
<style>
    .col{
        color:red;
    }
</style>

<div class="col">我是一段红色的文本 </div>
</body>
</html>
```

（3）通过 link 标签引入外联样式。

如下案例所示，运行之后会显示一段加粗的文本，加粗是在引入的外联样式 style.css 中设置。

```
<!DOCTYPE html>
<html lang="en">
<head>
    <meta charset="utf-8"/>
    <title></title>
    <link rel="stylesheet" href="style.css">
</head>
<body>
<div> 我是一段红色的文本 </div>
</body>
</html>
```

style.css 内容如下：

```
div{
font-weight: bold
}
```

在实际运用中,会将 3 种方式结合使用。一般会引入 2 个外联样式,1 个为公共样式,1 个为该文档自定义的样式。同时也会在元素标签中添加内联样式。

如下案例所示,运行之后将会显示一段字体颜色为红色、字体大小为 16px 的加粗文本。其中字体颜色是通过 style 标签设置,字体大小是通过内联样式设置,加粗是在引入的外部公共样式 style.css 中设置。

```
<!DOCTYPE html>
<html lang="en">
<head>
    <meta charset="utf-8"/>
    <title></title>
    <link rel="stylesheet" href="style.css">
</head>
<style>
    .col {
        color:red;
    }
</style>
<body>
<div style="font-size:16px" class="col">我是一段红色的文本</div>
</body>
</html>
```

style.css 内容如下:

```
div{
font-weight:bold;
}
```

以上 3 种方式的优先级分别为:内联样式最高,style 标签样式次之,外联样式最低。样式优先级指的是假设对一段文本,同时使用内联样式设置颜色为红色,使用 style 标签样式设置颜色为黄色,使用外联样式设置颜色为绿色,最终文本将显示为红色。如果没有内联样式，文本将显示为黄色；如果既没有内联样式，也没有 style 标签样式，文本将显示为绿色。

5.3 网页的交互——JavaScript

JavaScript 是一种运行在浏览器的脚本语言，主要功能是使网页具备一些更高级的动态效果并实现人机交互。比如在 Web 应用中最常见的轮播图功能，以及点击或者触摸设备后产生的响应。

如果把 HTML 比作人的躯体，那么 CSS 就是美丽的外衣，而 JavaScript 就赋予了人行动的能力。

5.3.1 JavaScript 历史

JavaScript 诞生于 1995 年。起初，这门语言的主要目的是处理以前由服务器端负责的一些表单验证。在那个绝大多数用户都在使用调制解调器上网的时代，用户填写完一个表单点击提交，需要等待几十秒，之后服务器反馈给用户时表示某个地方填错了，用户继续填写，再次验证……如果能在客户端完成一些基本的验证就能快速提高访问速度。当时走在技术革新前沿的网景（Netscape）公司，决定着手开发一种客户端语言，用来处理这种简单的验证。当时就职于 Netscape 公司的布兰登·艾奇（Brendan Eich）开始着手计划将 1995 年 2 月发布的 LiveScript 同时在浏览器和服务器中使用。为了赶在发布日期前完成 LiveScript 的开发，Netscape 与 Sun 公司成立了一个开发联盟。而此时，Netscape 公司为了搭上媒体热炒 Java 的"顺风车"，临时把 LiveScript 改名为 JavaScript，所以从本质上来说，JavaScript 和 Java 没什么关系。

1996 年 11 月，Netscape 公司决定将 JavaScript 提交给国际标准化组织 ECMA（European Computer Manufacturers Association，欧洲计算机制造商协会），希望这种语言能够成为国际标准。次年，ECMA 发布 262 号标准文件（ECMA-262）的第一版，规定了浏览器脚本语言的标准，并将这种语言称为 ECMAScript，这个版本就是 JavaScript 1.0。该标准从一开始就是针对 JavaScript 语言制定的，之所以不叫 JavaScript，有两个原因。其一是商标，Java 是 Sun 公司的商标，根据授权协议，只有 Netscape 公司可以合法地使用 JavaScript 这个名字，且 JavaScript 本身也已经被 Netscape 公司注册为商标。其二是想体现这门语言的制定者是 ECMA，而不是 Netscape，这样有利于保证这门语言的开放性和中立性。因此，ECMAScript 和 JavaScript 的关系是，前者是后者的规范，后者是前者的一种实现。

JavaScript 1.0 获得了巨大的成功，随后 Netscape 公司在 Netscape Navigator 3（网景浏览器）中发布了 JavaScript 1.1。之后作为竞争对手的微软在自家的 IE3 中加入了名为 JScript（名称不同是为了避免侵权）的 JavaScript 实现。而此时市面上意味着有 3 个不同的 JavaScript 版本，IE 的 JScript、Netscape 的 JavaScript 和 ScriptEase 的 CEnvi。当时还没有标准规定 JavaScript 的语法和特性。由于版本不同，暴露的问题日益加剧，JavaScript 的规范化最终被提上日程。

1997 年，以 JavaScript 1.1 为蓝本的建议被提交给了 ECMA，该协会指定 39 号技术委员会负责将其进行标准化，TC39 由各大公司以及其他关注脚本语言发展的公司的程序员组成，经过数月的努力完成了 ECMA-262——定义了一种名为 ECMAScript 的新脚本语言的标准。第二年，ISO/IEC（国标标准化组织和国际电工委员会）也采用了 ECMAScript 作为标准，即 ISO/IEC-16262。

1998 年 6 月，ECMAScript 2.0 发布。

1999 年 12 月，ECMAScript 3.0 发布，成为 JavaScript 的通行标准，得到了广泛支持。

2007 年 10 月，ECMAScript 4.0 草案发布，相当于对 3.0 版本做了大幅升级，并于次年 8 月发布正式版本。草案发布后，由于 4.0 版本的目标过于激进，各方对于是否通过这个标准发生了严重分歧。以 Yahoo、Microsoft、Google 为首的大公司，反对 JavaScript 的大幅升级，主张小幅改动；以 JavaScript 创造者 Brendan Eich 为首的公司，则坚持当前的草案。

2009 年 12 月，ECMAScript 5.0 正式发布。Harmony 项目则一分为二，一些较为可行的设想定名为 JavaScript.next 继续开发，后来演变成 ECMAScript 6；一些不是很成熟的设想，则被视为 JavaScript.next.next，在更远的将来再考虑推出。TC39 的总体考虑是，ECMAScript 5.0 与 ECMAScript 3.0 基本保持兼容。较大的语法修正和新功能加入，将由 JavaScript.next 完成。当时，JavaScript.next 指的是 ECMAScript 6。第六版发布以后，将指 ECMAScript 7。TC39 预计，ECMAScript 5.0 会在 2013 年的年中成为 JavaScript 开发的主流标准，并在此后 5 年中一直保持这个位置。

2011 年 6 月，ECMAscript 5.1 发布，并且成为 ISO 国际标准（ISO/IEC 16262: 2011）。到了 2012 年年底，所有主流浏览器都支持 ECMAScript 5.1 的全部功能。

2013 年 3 月，ECMAScript 6 草案冻结，不再添加新功能。新的功能设想将被放到 ECMAScript 7。

2013 年 12 月，ECMAScript 6 草案发布。然后是 12 个月的讨论期，听取各方反馈。

2015 年 6 月，ECMAScript 6 正式发布，并且更名为"ECMAScript 2015"。这是因为 TC39 委员会计划，每年发布一个 ECMAScirpt 版本。

5.3.2　JavaScript 语法

1. 数据类型

JavaScript 数据类型分为基本类型和引用类型。

基本类型有数字（Number）、字符串（String）、布尔（Boolean）、未定义（Undefined）、空（Null）、Symbol（ES6 语法新增）6 种。

引用类型有对象（Object）、数组（Array）、函数（Function）3 种。

typeof 操作符：在介绍数据类型之前，先介绍一下 typeof 操作符。其作用就是检测数据类型，返回值是表示某个数据类型的字符串。

Number 类型：表示数字，包含了所有的数值类型，在 JavaScript 中没有像高级语言那样细分成 int、double、deciable。它的取值范围是 $-1.7976931348623157E+308$ ~ $1.7976931348623157E+308$，就是 $-1.7976931348623157 \times 10^{308}$ ~ $1.7976931348623157 \times 10^{308}$。Number 类型有一个特殊的值 NaN，它是 Number 类型中一个非数值的值，就像无穷大，所以 NaN 不等于 NaN。

如：`let x = 1;//x 为 Number 类型`

String 类型：表示字符串，String 类型是 JavaScript 中一个比较重要的类型。String 类型就是一串字符。字符串被单引号或双引号包围，如下所示。

```
let str = "hello world";    // 可以使用双引号
let str = 'hello world';    // 可以使用单引号
```

Boolean 类型：Boolean 类型表示值为 true 或者 false 的数据类型，需要注意的是，Boolean 类型和 Number 类型可以相互转换，比如 true 可以转化为数值 1，false 可以转化为数值 0，如下所示。

```
let x = false;
```

```
let y = true;
```

Undefined 类型：这是全局对象的一个特殊属性，顾名思义就是未定义。Undefined 类型只有一个值为 undefined，表示一个变量定义了但未赋值。这个值是为了区分空对象指针与未经初始化的变量。

如：

```
let a;
console.log(a); // 返回 undefined
```

下面是几种常见的会出现 undefined 的场景。

① 获取一个对象的属性（原始属性和 prototype 扩展的属性）不存在时，返回 undefined。

```
let obj={
    name:'jack'
};
console.log(obj.age);// 返回 undefined
```

② 当一个函数 function（）{} 没有明确的返回值时，返回 undefined。

```
function empty(){};
console.log(empty());// 返回 undefined
```

③ 当一个函数中的参数有多个形参，调用时，实参数量小于形参数量，那么其他的参数返回值就为 undefined。

```
function fn(x,y) {
    console.log(x,y);
};
fn(1)// 返回 1 和 undefined
```

Null 类型：表示为空或者不存在的对象引用，Null 类型也只有一个 null 值。

```
let obj=null;
```

```
console.log(obj)// 返回 null
```

值得注意的是，JavaScript 中的 if 判断语句无论是 Null 类型还是 Undefined 类型，它们的返回值都是 false。

```
if (!null){
    console.log("1");//1
}
if (!undefined){
    console.log("2");//2
}
```

Symbol 类型：ES6 引入了一种新的原始数据类型 Symbol，表示独一无二的值。Symbol 值通过 Symbol 函数生成。这就是说，对象的属性名现在可以有两种类型，一种是原来就有的字符串，另一种就是新增的 Symbol 类型。凡是属性名属于 Symbol 类型，就都是独一无二的，可以保证不会与其他属性名产生冲突。

```
let s = Symbol();
console.log(typeof s);// "symbol"
```

上面代码中，变量 s 就是一个独一无二的值。typeof 运算符的结果，表明变量 s 是 Symbol 类型，而不是字符串或者其他类型。

2. 变量提升

在 JavaScript 中，函数及变量的声明都将被提升到函数的最顶部。变量可以在使用后声明，也就是变量可以先使用再声明。

```
x = 5;
console.log(x)//5
var x;
```

以上代码会输出 5，x 会被提升到最顶部。

3. 作用域

作用域是变量与函数的可访问范围，其控制变量和函数的可见性和生命周期。在 JavaScript 中，对象和函数同样也是变量。作用域分为局部作用域和全局作用域。

在函数内声明的变量是局部变量，只能在函数内部访问。局部变量在函数开始执行时创建，函数执行完后局部变量会自动销毁。

在函数外声明的变量为全局变量，全局变量拥有全局作用域，所有函数和脚本都可以使用。全局变量在页面关闭后销毁。

当声明一个函数时，局部作用域一级一级向上用花括号 {} 括起来，就形成了作用域链。在作用域链中，执行函数时，总是先查找函数内部的变量，如果内部没有，则查找最近的父作用域，没有父作用域时，就会去全局作用域查找。

```
let x = 5;
function fn(){
    x = 10;
    return x;
};
console.log(fn())//10
```

上面代码中执行 fn 函数时，先在其内部查找是否有 x 变量，如果有，就取内部变量值 10；如果没有，则在全局作用域查找。

4. 循环

循环表示可多次执行相同的代码块。JavaScript 支持不同类型的循环。

for，多次遍历代码块。

for...in，遍历对象属性。

while，当指定条件为 true 时，循环一段代码块。

do...while，当指定条件为 true 时，循环一段代码块。

for 循环示例：

```
for (let i = 0; i < 3; i++) {
    console.log(i);
}
```

输出 0，1，2。

while 循环示例：

```
let i=0;
while(i<3){
    i++;
```

```
        console.log(1)
    };
```

输出 0，1，2。

do...while 循环示例：

```
let i = 0;
    do{
//do 中是要执行的代码
        console.log(i);
        i++;
    }while(i<3)//while 中是满足的条件
```

输出 0，1，2。

在循环中有两个关键字 break 和 continue。break 表示跳出循环。continue 表示跳过满足条件的语句。

break 示例如下：

```
for (let  i = 0; i < 3; i++) {
    if(i == 0){
        console.log(i);
        break;
    }
}
```

输出 0。该循环中使用了 break 关键字，满足 i == 0 之后就不会再执行。

continue 示例如下：

```
for (let  i = 0; i < 3; i++) {
    if(i == 0){
        continue;
    }
    console.log(i);
}
```

5. 原型链与继承

JavaScript 所有对象都可以从原型对象上继承属性和方法。每个构造函数都有一个原型对象，原型对象包括一个指向构造函数的指针。而实例中包含一个指向原型对象的内部指针。即构造函数的原型对象 prototype 等于实例对象的 __proto__。

假设构造函数的原型对象等于另一个构造函数的实例对象，那么实例对象的 __proto__ 就拥有另一个构造函数的属性和方法。而所有函数的顶层都是 Object 函数。所以另一个构造函数的实例对象又拥有 Object 函数原型对象的属性和方法，以此构成一个链条就叫作原型链。实例对象可以拥有原型对象上的属性和方法就叫作继承。

```
function fn() {
    this.name="jack";
};
function fn1() {
    this.age=18;
}
fn.prototype = new fn1();
let x= new fn();
console.log(x.name);//jack
console.log(x.age);//18, age 属性是继承自构造函数 fn1 的原型对象
console.log(x.__proto__ === fn.prototype);//true
// x 是 fn 的实例对象，所以实例对象的内部指针指向构造函数的原型对象
console.log(fn.prototype.__proto__ === fn1.prototype);//true
//fn 的原型对象等于 fn1 的实例对象，所以原型对象的指针指向 fn1 的原型对象
console.log(fn1.prototype.__proto__ === Object.prototype);//truc
// 原型对象的指针指向顶层函数 Object 的原型对象
```

综上，最后 fn 的实例对象 x 的指针指向顶层函数 Object 的原型对象，因此 x 既拥有自身构造函数 fn 上的属性和方法，也拥有 fn1 构造函数的属性和方法以及 Object 构造函数的属性和方法，即

```
console.log(x.__proto__.__proto__.__proto__ === Object.prototype);//true
```

这里介绍一下 instanceof 操作符，instanceof 操作等是用来判断一个对象是否是构造函数的实例对象。

```
console.log(x instanceof fn);//true
//x 是 fn 的实例对象
console.log(x instanceof fn1); //true
//x 是 fn 的实例对象，而 fn 的原型对象等于 fn1 的实例对象，所以 x 也属于 fn1 的实例
```

对象

```
console.log(fn instanceof Object); //true
// 任何函数都可以继承顶层构造函数 Object 的属性和方法
console.log(fn1 instanceof Object);
// 任何函数都可以继承顶层构造函数 Object 的属性和方法

let arr = new Array();
console.log(arr instanceof Array);//true
//arr 是 Array 函数的实例对象
console.log(arr instanceof Object);//true
//Array是 Object 的子集，所以 arr 也属于 Object
```

6. 闭包

闭包是指在有权访问另一个函数作用域中变量的函数。

```
function fn() {
    let x = 10;
    return function () {
        return x;
    }
};
var result = fn();
console.log(result());//10
```

在上文的作用域中讲到，在函数内定义的变量为局部变量，只能在作用域内访问。因此正常情况下在全局环境下无法访问 fn 函数作用域内的变量 x。但是利用闭包的特性，在父函数 fn 中返回一个局部匿名函数，该匿名函数可访问 fn 中的变量 x。在全局环境下，先调用 fn() 返回一个内部匿名函数，再调用内部匿名函数，即可得到 x 的值为 10。

5.3.3 AJAX 介绍

AJAX 全称为 Asynchronous JavaScript And XML，意为异步 JavaScript 和 XML。AJAX 在不重新加载整个页面的情况下，可以与服务器交换数据并更新部分网页内容。

AJAX 的使用分为以下几个步骤。

（1）创建 XMLHttpRequest 对象，代码如下。

```
var xmlhttp;
if (window.XMLHttpRequest) {
    // IE7+、Firefox、Chrome、Opera、Safari 执行代码
    xmlhttp=new XMLHttpRequest();
} else {
    // IE6、IE5 执行代码
    xmlhttp=new ActiveXObject("Microsoft.XMLHTTP");
}
```

（2）向服务器发送请求，代码如下。

```
xmlhttp.open("GET","ajax_info.txt",true);
xmlhttp.send();
```

open 方法中第 1 个参数为请求的类型（GET 或 POST），第 2 个参数为 URL（请求接口地址），第 3 个参数为是否异步处理请求（true 为异步，false 为同步）。

send 方法是将请求发送到服务器。

（3）onreadystatechange 事件。

当请求被发送到服务器时，需要执行一些基于响应的任务。每当 readyState 改变时，就会触发 onreadystatechange 事件。

```
xmlhttp.onreadystatechange=function(){
    if (xmlhttp.readyState==4 && xmlhttp.status==200){
        console.log(xmlhttp.responseText);// 请求成功之后的返回值
    }
}
```

readyState 属性存有 XMLHttpRequest 的状态，并从 0 到 4 发生如下变化。

0：请求未初始化。

1：服务器连接已建立。

2：请求已接收。

3：请求处理中。

4:请求已完成,且响应已就绪。

status 属性是 http 状态码,常用的 200 表示请求成功,404 表示找不到网页,500 一般表示服务端出错。

5.4 Web 网页案例

5.4.1 案例说明

开发网页的工具有很多,有轻量极的 Notepad++,适合初学者使用;有功能渐进式的 VSCode,以及功能强大的 HBuilde 和 WebStorm,这些工具直接在官网下载并安装即可。本项目使用的开发工具为 WebStorm。

5.4.2 案例代码

如果自己制作一个简单的网页,操作步骤如下。

第一步:制作一个只有 HTML 的网页。

```html
<!DOCTYPE html>
<html lang="en">
<head>
    <meta charset="UTF-8">
    <title>Title</title>
</head>
<body>
<div>
    <div> 再别康桥 </div>
    <div>-- 徐志摩 </div>
    <div> 轻轻的我走了,正如我轻轻的来;</div>
    <div> 我轻轻的招手,作别西天的云彩。</div>
    <div> 那河畔的金柳,是夕阳中的新娘;</div>
    <div> 波光里的艳影,在我的心头荡漾。</div>
     <div> 软泥上的青荇,油油的在水底招摇;</div>
    <div> 那榆荫下的一潭,不是清泉,是天上虹;</div>
    <div> 揉碎在浮藻间,沉淀着彩虹似的梦。</div>
```

```
        <div>寻梦？撑一支长篙，向青草更青处漫溯；</div>
        <div>满载一船星辉，在星辉斑斓里放歌。</div>
        <div>但我不能放歌，悄悄是别离的笙箫；</div>
        <div>夏虫也为我沉默，沉默是今晚的康桥！</div>
        <div>悄悄的我走了，正如我悄悄的来；</div>
        <div>我挥一挥衣袖，不带走一片云彩。</div>
    </div>
</body>
</html>
```

以上只有 HTML 代码的内容，也是一个最简单的网页示例，在浏览器显示效果如图 5-1 所示。

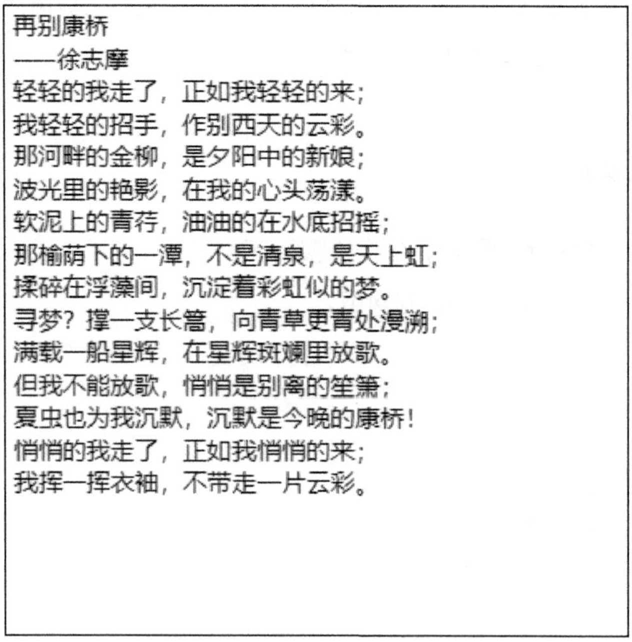

```
再别康桥
——徐志摩
轻轻的我走了，正如我轻轻的来；
我轻轻的招手，作别西天的云彩。
那河畔的金柳，是夕阳中的新娘；
波光里的艳影，在我的心头荡漾。
软泥上的青荇，油油的在水底招摇；
那榆荫下的一潭，不是清泉，是天上虹；
揉碎在浮藻间，沉淀着彩虹似的梦。
寻梦？撑一支长篙，向青草更青处漫溯；
满载一船星辉，在星辉斑斓里放歌。
但我不能放歌，悄悄是别离的笙箫；
夏虫也为我沉默，沉默是今晚的康桥！
悄悄的我走了，正如我悄悄的来；
我挥一挥衣袖，不带走一片云彩。
```

图 5-1

第二步：使网页穿上"美丽的外衣"，添加 CSS 代码。

以上的网页没有任何布局和色彩。下面加入 CSS 代码来格式化，也就是美化网页。

```
<!DOCTYPE html>
<html lang="en">
<head>
    <meta charset="UTF-8">
```

```html
        <meta name="viewport" content="width=device-width, initial-scale=1.0, minimum-scale=1.0, maximum-scale=1.0, user-scalable=no" />
        <title>Title</title>
    </head>
    <style>
        body{
            background:#7976072b;
        }
        div{
            text-align:center;
            margin-top:20px;
            margin-bottom:20px;
            font-size:18px;
            font-family:cursive;
            color:#345c05;
        }
        .title{
            font-weight:bold;
            font-size:20px;
        }
        .author{
            margin-left:200px;
        }
    </style>
    <body>
        <div class="title"> 再别康桥 </div>
        <div class="author">——徐志摩 </div>
        <div> 轻轻的我走了，正如我轻轻的来；</div>
        <div> 我轻轻的招手，作别西天的云彩。</div>
        <div> 那河畔的金柳，是夕阳中的新娘；</div>
        <div> 波光里的艳影，在我的心头荡漾。</div>
        <div> 软泥上的青荇，油油的在水底招摇；</div>
        <div> 那榆荫下的一潭，不是清泉，是天上虹；</div>
        <div> 揉碎在浮藻间，沉淀着彩虹似的梦。</div>
        <div> 寻梦？撑一支长篙，向青草更青处漫溯；</div>
        <div> 满载一船星辉，在星辉斑斓里放歌。</div>
        <div> 但我不能放歌，悄悄是别离的笙箫；</div>
        <div> 夏虫也为我沉默，沉默是今晚的康桥！</div>
        <div> 悄悄的我走了，正如我悄悄的来；</div>
        <div> 我挥一挥衣袖，不带走一片云彩。</div>
    </body>
```

```
</html>
```

效果如图 5-2 所示。

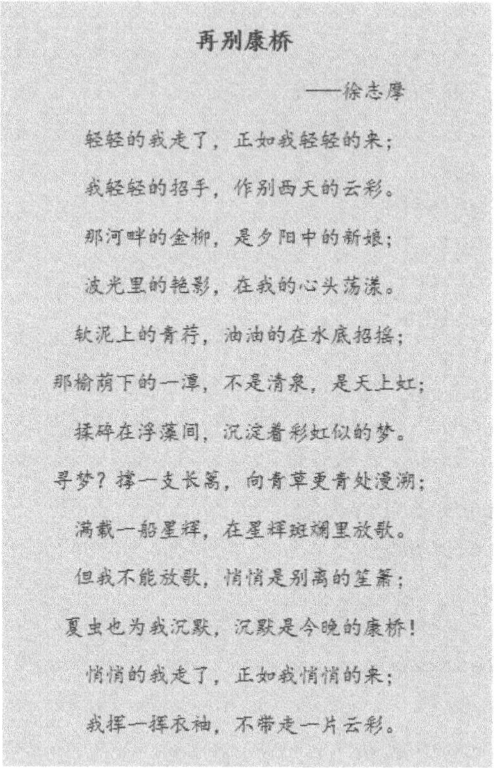

图 5-2

图 5-1 和图 5-2 充分体现了是否使用 CSS 的区别。而这只是小试牛刀，所有精美的网页，其布局和色彩的实现都得益于强大的 CSS。

第三步：加入 JavaScript。

使用 HTML+CSS 制作出来的网页是纯静态的，没有任何动态效果，也无法与用户进行交互。假设要实现如下功能。

页面有一个按钮【点击查看再别康桥】。用户点击按钮之后，这首诗会出现，并且在网页上旋转一圈。

用户点击按钮之后这首诗才出现，这种行为叫做网页与用户交互。这首诗在页面旋转一圈，是一个简单的动画效果。要实现这个功能 JavaScript 就派上用场了。

```
<!DOCTYPE html>
```

```html
<html lang="en">
<head>
    <meta charset="UTF-8">
    <meta name="viewport" content="width=device-width, initial-scale=1.0, minimum-scale=1.0, maximum-scale=1.0, user-scalable=no" />
    <title>Title</title>
</head>
<style>
    body{
        background:#7976072b;
    }
    div{
        text-align:center;
        margin-top:20px;
        margin-bottom:20px;
        font-size:18px;
        font-family:cursive;
        color:#345c05;
    }
    .title{
        font-weight:bold;
        font-size:20px;
    }
    .author{
        margin-left:200px;
    }
    .hide{
        display:none;
    }
</style>
<body>
<div id="btn">
    <button>
        点击查看再别康桥
    </button>
</div>
<div class="hide" id="poem">
<div class="title">再别康桥 </div>
<div class="author">-- 徐志摩 </div>
<div> 轻轻的我走了，正如我轻轻的来； </div>
<div> 我轻轻的招手，作别西天的云彩。 </div>
```

```html
        <div>那河畔的金柳，是夕阳中的新娘；</div>
        <div>波光里的艳影，在我的心头荡漾。</div>
        <div>软泥上的青荇，油油的在水底招摇；</div>
        <div>那榆荫下的一潭，不是清泉，是天上虹；</div>
        <div>揉碎在浮藻间，沉淀着彩虹似的梦。</div>
        <div>寻梦？撑一支长篙，向青草更青处漫溯；</div>
        <div>满载一船星辉，在星辉斑斓里放歌。</div>
        <div>但我不能放歌，悄悄是别离的笙箫；</div>
        <div>夏虫也为我沉默，沉默是今晚的康桥！</div>
        <div>悄悄的我走了，正如我悄悄的来；</div>
        <div>我挥一挥衣袖，不带走一片云彩。</div>
    </div>
    <script>
        var btn = document.getElementById("btn");
        var poem = document.getElementById("poem");
        btn.onclick = function () {
            // 点击后消除btn按钮
            btn.style.display = "none";
            // 点击后出现诗歌
            poem.style.display = "block";
            // 配置参数
            var index = 0,degreeStart = 0,degreeEnd = 360,speed = 0.03,angle = 10,time = 10;
            // 开启定时器
            var timer = setInterval(()=>{
                // 当旋转角度大于或等于360度时停止定时器
                if(degreeStart>=degreeEnd){
                    // 消除定时器
                    clearInterval(timer);
                    timer = null;
                    return;
                }
                index += speed;
                degreeStart += angle;
                poem.style.transform = "rotateZ("+degreeStart+"deg)"
            },time)
        }
    </script>
</body>
</html>
```

因为是动态效果，这里就不提供截图了。

本小节通过一个很简单的案例展现出 HTML、CSS、JavaScript 各自的分工。在目前的市场环境下，只要是在浏览器中运行的网页，所有的页面内容、页面样式以及页面动效与交互，都是由 HTML、CSS、JavaScript 来实现的。

第 6 章 前端主流框架——React

【本章导读】

◎ React 主要特点及生命周期

◎ React 开发环境搭建

◎ 编写 React 案例，对比 React 框架开发的网页与原生网页的区别

6.1 React 概述

React 是由 FaceBook 内部团队开发的一个前端框架，于 2013 年 5 月开源，是目前市场上主流的前端开发框架。在此之前，jQuery 框架一直是前端框架的首选，React 问世后，其单页面应用、虚拟 DON、高性能等特点，可以说是对整个前端领域的颠覆。

6.1.1 React 主要特点

（1）单页面：传统的 Web 项目中，用户可浏览的页面有几个，在开发的时候就有几个 HTML 文件。比如用户浏览首页，会有一个 index.html 页面；用户浏览详情页，会有一个 detail.html 页面；用户浏览个人中心页面，会有一个 personal.html 页面。但是在 React 中，不管是多大型的项目，有且仅有一个页面。在 React 中所提及的页面，都是由一个个组件构成，其实现逻辑是由 JS 动态生成，仅有的一个页面只不过是一个容器。

（2）组件化：组件可以理解为一个页面，或者一个功能模块，甚至一个页面元素。通俗地讲，可以把组件理解为一个物品的零件。比如一台计算机是由显示屏、内存条、键盘等组成。那么在网页里面，一个功能零件就是一个组件。组件的颗粒度要尽可能的小，即功能单一化，能最大程度地实现高可复用。相似功能或者相似页面的内容，可以写成一个通用的组件，这将大大减少代码量，使逻辑结构更加清晰。就像一台计算机的每个零件

都可以被拆解，这样方便维修。假设计算机的显示屏坏了，维修人员可以拿相同型号的显示屏修好计算机。试想如果计算机的安装是全部焊接固定，那么一旦出现一点问题，计算机就直接报废。代码的设计也是一样的，组件化的设计就是充分体现低耦合。

（3）高性能：频繁地操作 DOM 节点会导致页面重绘或者回流，这将会增加页面加载时间，影响用户访问速度。而在 React 中采用的是虚拟 DOM 原理，即通过 JSX 语法绘制出来的元素只是一种类似 DOM 的数据结构，并不是真正的 DOM。这种原理大大减小了 DOM 节点操作频率，优化了性能。

（4）单向数据流：在 React 中数据流是单向的，数据通过组件的 props 和 state 层层向下传递，如果要添加反向数据流，则需要通过父组件将回调函数传递给子组件。每当状态更新时，回调就会触发，父组件会调用 setState，以此来重新渲染页面。

6.1.2　React 生命周期

React 生命周期是指组件在运行过程中会经历的初始化、更新、卸载 3 个阶段，而每个阶段又有不同的生命周期，在 React 生命周期内可以通过钩子函数来完成不同的业务逻辑。

1. 初始化阶段

constructor：此函数用来做一些组件的初始化工作，比如定义 state 和 props。

componentWillMount：此函数是在组件挂载到 DOM 节点前调用，可以在这个阶段修改初始定义的 state 和 props。

render：页面渲染。

componentDidMount：此函数是在组件挂载到 DOM 节点后调用。在此阶段可调取服务端接口数据，更新组件，也可以获取真实的 DOM 节点。这是 React 中最常用的生命周期。

2. 更新阶段

componentWillReceiveProps：此函数表示父组件渲染时，子组件也更新，可通过此函数来监听 props 和 state 的变化。

shouldComponentUpdate：此函数可以判断当前 state 是否发生变化，如果没有发生变化，可在此返回 false，则不会重新渲染页面。因此在此期间可以做性能优化。

componentWillUpdate：此函数在更新页面前调用。

render：根据改变的 state 和 props 重新渲染页面。

componentDidUpdate：此函数是组件已经完成更新，新的虚拟 DOM 节点已经完成挂载。

3. 卸载阶段

componentWillUnmount：此函数在组件被卸载前调用，可在此清除计时器及取消事件监听，还可以移除手动创建的 DOM 节点，避免内存泄漏。

6.2 React 开发环境搭建

目前市面上主流的前端开发模式都具有工程化、模块化、系统化的特点。所有的应用模块会放置于 NPM 包中统一下载使用，而 NPM 包是 Node 下面的一个子包，所以前端程序的运行需要在 Node 环境下进行。

6.2.1 Node 安装

（1）在 Node 官网下载符合计算机配置的 Node 版本，如图 6-1 所示。

（2）设置 Node 环境变量，以 Win10 为例。

① 右击"此电脑"，在快捷菜单中选择【属性】，如图 6-2 所示。

② 在弹出的【系统】窗口中点击【高级系统设置】，在弹出的【系统属性】对话框中点击【环境变量】，如图 6-3 所示。

图 6-1

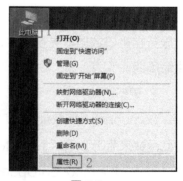

图 6-2

③ 在弹出的【环境变量】对话框中双击【Path】一栏，如图 6-4 所示。

④ 在弹出的【编辑环境变量】对话框中点击【新建】，在空白处将 Node 安装路径

添加进去，点击【确定】，如图 6-5 所示。

⑤ 点击"环境变量"对话框中的【确定】。

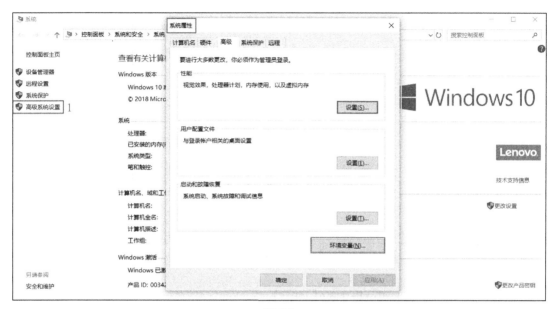

图 6-3

图 6-4

图 6-5

⑥ 点击【系统属性】对话框中的【确定】。

⑦ 关闭【系统】窗口。

值得注意的是，下载完 Node 之后，也会自动下载 npm 包模块。NPM 是前端包管理工具，在前端工程化开发中必不可少。它的功能主要有以下两种。

① 开发者可以从 NPM 服务器将第三方模块下载到本地使用。

命令为 npm install 模块名称。

② 开发者也可以将自己编写的模块上传到 NPM 服务器供他人下载使用。

命令为 npm publish。前提是注册并登录了 NPM 账号。

6.2.2 React 项目构建

（1）在桌面新建文件夹"secondary"，用 WebStorm 编辑器打开。

（2）执行 npm install create-react-app 命令安装脚手架工具，如图 6-6 所示。

（3）执行 create-react-app secondary 命令安装 React 项目脚手架，通过 create-react-app 模块直接初始化项目，执行完之后将会下载一个完整的 React 脚手架结构，附带一些简单的代码示例，如图 6-7 所示。

（4）执行 cd secondary 命令进入项目根目录，执行 npm run start 启动项目，项目启动成功之后在控制台日志中可以看到访问地址，如图 6-8 所示。

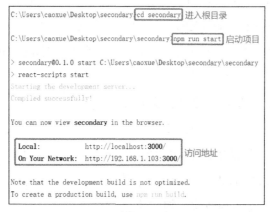

图 6-6

图 6-7

图 6-8

（5）在浏览器地址栏输入 http：//localhost：3000/ 打开项目，看到如图 6-9 所示的界面，表示 React 开发环境搭建成功。注意端口号可能根据计算机运行程序有所调整，并不一定是 3000，具体的端口号以控制台日志中的访问地址为准。

（6）初始化项目目录，如图 6-10 所示。

图 6-9

图 6-10

secondary 是最开始创建的文件夹,里面有 3 部分内容,分别为 node_modules、secondary、package-lock.json。

① node_modules 是使用 npm install create-react-app 命令下载的模块。

② secondary 是使用 create-react-app secondary 命令下载的项目根路径。

③ package-lock.json 是下载 create-react-app 模块时自动生成的描述文件,记录了整个 node_modules 文件夹的描述信息,呈现为一个树状结构。它包含的内容有模块版本信息、每个模块所依赖的其他模块信息以及下载地址等。

第二个 secondary 才是整个项目的根路径,来看看它内部具体包含了哪些内容。secondary 目录如图 6-11 所示。

secondary 根目录下包含的内容有 node_modules、public、src 以及一些描述文件。

① node_modules 是初始化项目时所用到的模块包文件,一般执行 npm install 命令就会下载这个模块包文件,如果下载具体的模块,将会更新这个文件夹。

② public 中有 favicon.ico 图片文件、index.html 网页文件以及 manifest.json 描述文件(包含项目的显示名称、图标、主题颜色等)。

③ src 是整个项目的核心文件,其中 App.css

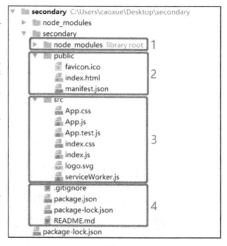

图 6-11

是 App 组件的样式编写文件，App.js 是页面和业务逻辑编写文件，App.test.js 是编写测试用例的文件，index.css 是通用样式编写文件，index.js 是项目入口文件，logo.svg 是用 svg 绘制的图片文件，serviceWorker.js 是离线配置文件（也就是断网之后依然可以访问）。

④ 这一部分也是一些配置和描述文件，.gitignore 文件是 git 版本控制的配置文件；package.json 是本项目描述文件，比如项目名称、项目版本、项目所需要的依赖模块等；package-lock.json 与 package.json 有些类似，但它主要包含项目依赖的模块信息描述；README.md 文件是项目中执行的命令功能描述，比如 npm start 命令是启动项目、npm test 命令是启动测试、npm run build 命令是打包项目等。

（7）在以上脚手架项目的基础上做以下目录的调整。

① 在 src 目录下新建 pages 文件夹，用于存放业务模块文件。

② 在 src 目录下新建 image 文件夹，用于存放图片资源文件。

③ 在 src 目录下新建 model 文件夹，用于存放数据模型，也就是前端页面模拟的接口返回 json 格式的数据。因为在正常前、后端分离模式下，前端与后端同步开发，所以前端页面编写完成之后，后端接口数据不一定编写完成，这就需要前端自己模拟数据去填充页面。

④ 在 src 目录下新建 common.css 文件，用于编写公共样式。

⑤ 在 src 目录下新建 router.js 文件，用来配置路由，实现页面访问与跳转。

⑥ 将 src 目录下的 App.css、App.test.js、index.css、logo.svg 都删除，并将 index.js 引用的 index.css 替换成 common.css。

6.3　React 案例

在前面搭建的 React 脚手架中已经有实例了，现在用 React 框架来编写上文原生网页（指使用 HTML/CSS/JavaScript 编写的网页，不涉及任何框架）示例中的那首诗《再别康桥》。

6.3.1　案例代码

App.js 的内容如下：

```jsx
import React from 'react';
import Login from './pages/login/Login';
import Router from './router';
import './App.css';

export default class App extends React.Component {
    componentDidMount() {
        var index = 2;
        var timer = setInterval(function () {
            let num = document.getElementById("num");
            num.innerHTML = index;
            index--;
            if (index < 0) {
                let content = document.getElementById("content");
                num.style.display = 'none';
                content.style.display = 'block';
                clearInterval(timer)
            }
        }, 1000)
    }

    render() {
        return (
            <div>
                <div id="content">
                    <div style={{fontWeight:"bold", fontSize:"20px"}}>再别康桥</div>
                    <div style={{marginLeft:"200px"}}>——徐志摩</div>
                    <div>轻轻的我走了，正如我轻轻的来；</div>
                    <div>我轻轻的招手，作别西天的云彩。</div>
                    <div>那河畔的金柳，是夕阳中的新娘；</div>
                    <div>波光里的艳影，在我的心头荡漾。</div>
                    <div>软泥上的青荇，油油的在水底招摇；</div>
                    <div>那榆荫下的一潭，不是清泉，是天上虹；</div>
                    <div>揉碎在浮藻间，沉淀着彩虹似的梦。</div>
                    <div>寻梦？撑一支长篙，向青草更青处漫溯；</div>
                    <div>满载一船星辉，在星辉斑斓里放歌。</div>
                    <div>但我不能放歌，悄悄是别离的笙箫；</div>
                    <div>夏虫也为我沉默，沉默是今晚的康桥！</div>
                    <div>悄悄的我走了，正如我悄悄的来；</div>
                    <div>我挥一挥衣袖，不带走一片云彩。</div>
```

```
            </div>
            <div id="num">
                3
            </div>
        </div>
    );
  }
}
```

6.3.2 React 框架与原生网页的区别

1. 模块化开发

从前端到大前端的跨越，工程化、模块化开发是最为显著的特征。工程化指的是通过安装、命令等方式自动构建工程，并部署到本地服务器。模块化开发指的是每个组件都可以视为一个模块，而模块与模块之间的依赖就靠导入（关键字 import）与导出（关键字 export）。导入指的是需要用到别的模块时，首先将模块导入进来；导出指的是将模块暴露出去，供其他模块使用。在 React 中，组件的最前面就是通过 import 导入的各种依赖模块。而模块，或者组件本身也会通过 export 关键字对外暴露。

2. export 与 export default

使用 import 关键字导入依赖时，有两种表达形式，一种是以上代码中示范的 import React from 'react'，还有一种可以使用 import {React} from 'react' 这种形式。那么什么情况下需要使用 {} 呢？这取决于模块在对外暴露时是使用 export 还是 export default。这两者又有什么区别呢？其实很简单，在一个 js 文件中，如果组件或者函数使用 export 对外暴露，则在引入的时候需要使用 {}。这说明一个 js 文件中有多个模块对外提供，即一个 export 对应的就是一个模块。而使用 export default 时，则表示一个 js 文件中有且仅有一个默认的模块对外提供，所以在引入的时候就不需要使用 {}。

那么模块在对外提供的时候，什么时候用 export，什么时候用 export default 呢？这个没有严格规定，主要还是根据业务场景来判断。一般独立的组件是使用 export default，而同一个 js 文件中需要对外暴露多个方法则可以使用 export。具体的应用详见后面的实际项目。

3. class 语法

JavaScript 是一个弱类型语言，因此在 JavaScript 的历史中，一直都没有类的概念，class 是 ES6 的语法，表示类。从简单的层面去理解，一个类其实就是一个构造函数。而在 React 中，一个类就表示一个组件。其实不管是构造函数，还是类，抑或是组件，只是叫法不一样，它们的本质都是数据结构的一种抽象，最终都会生成一个实例对象。class 后面紧跟着组件名称，需要注意的是组件名称首字母要大写。

4. extends 继承

extends 顾名思义就是继承，React.Component 可以理解为所有自定义组件的父组件，该组件提供了 props 属性、context 属性、refs 属性等，以及更新组件最关键的 setState 方法。只有使用了 extends 关键字，才可以在自定义组件中继承这些属性和方法。

5. 生命周期

在 6.1.2 节中有介绍 React 的生命周期。在本案例中，正好使用了最常用的 componentDidMount 函数。可以看到，在本案例中有获取 DOM 节点的操作，而正好在 componentDidMount 生命周期中可以获取真实的 DOM 节点。其次，本案例中使用了 render 函数。该函数是一个纯函数，返回的是由 JSX 语法渲染出来的页面元素。

6. JSX 语法

6.3 节的案例中被 render 函数 return 的所有节点看作 HTML 语法，却不是在 html 文件中编写。这种语法结构被称作 JSX 语法，该语法是 JavaScript 的一种拓展语法，在 React 中用来描述用户界面。在 JSX 语法中，所有的表达式要包含在 {} 中，而 JSX 本身也是一种表达式，这就意味着，在其中可以使用 if 判断或者 for 循环。使用 JSX 语法编写的内容经过编译之后，会转换为普通的 JavaScript 对象。

第 7 章　前端常用开发工具/库

【本章导读】

◎ 打包工具之 webpack

◎ 页面跳转之 react-router-dom

◎ 前端组件库之 ant-design

◎ CSS 预处理器之 Less

7.1　打包工具之 webpack

webpack 是目前前端开发中最流行的打包工具之一，是 JavaScript 的静态模块打包器。在打包的过程中，webpack 会根据业务逻辑构建一个依赖关系图。每一个依赖的单元都是一个模块，这个模块可以是 js，也可以是图片资源或者 CSS 资源。在使用 webpack 时需要安装 webpack（打包工具）、webpack-cli（命令行工具）、webpack-dev-server（Node 服务器）3 个基础模块。

7.1.1　webpack 核心原理

1. 一切皆模块

模块不仅包括 js 文件，还包括 css 文件和图片文件。在 webpack 的世界里，这些都可以是一个模块，一般通过 require 或者 import 的方式引入资源。前者属于 CommonJS 规范，后者属于 ES6 语法。

2. 按需加载

单页面应用在初始时要加载非常大的代码量，由于加载时间过长会导致页面出现白屏或者等待现象。按需加载指的是随着用户的操作，每次只加载当前页面或业务功能所需要的代码块。webpack 内置了非常强大的代码分割功能，可以实现按需加载。

7.1.2 webpack 核心概念

1. 入口（entry）

打包的入口点，也就是关系依赖图的起点。在 webpack 中通过 entry 属性来配置。

2. 出口（output）

打包的出口，用于指定打包出来之后的路径、文件名。

3. loader 编译器

loader 是除了 js 文件之外的其他文件的编译器。上文提到在 webpack 的世界里，一切皆模块。但 webpack 只能识别 js 文件，其他文件需要通过 loader 编译器转换。loader 在 webpack 的 module 属性中配置。其中 rules 配置编译规则，test 用于正则匹配，exclude 用于排除特定条件，use-loader 是 test 匹配到的解析器模块，use-options 主要与 use-loader 配合使用。

4. 插件（plugins）

如果说 loader 主要是用于转换文件，那么插件则可以用来执行范围更广的任务。比如将 ES5 解析成 ES6，LESS 文件解析成 CSS 文件，为了兼容浏览器。plugins 是将 loader 之后的文件进行优化分类、压缩并提供公共代码等。

5. 打包模式（mode）

除了以上 4 大核心概念之外，webpack 中还有模式切换配置。在 mode 属性中进行配置时，mode 提供了两种打包模式，一种是开发环境，命令为 webpack --mode = development，打包出来的文件未经过压缩；另一种是生产环境，命令为 webpack --mode = production，打包出来的文件是经过压缩的。

6. 服务器配置（devServer）

服务器配置指的是配置本地服务器，可以配置启动端口、主机地址、是否热启动等信息。

7.1.3 webpack 参数配置说明

在项目中配置 webpack 需要在根目录下创建 webpack.config.js 配置文件，通过这个文件对 webpack 进行配置。

```
var path = require('path');
```

```javascript
var ExtractTextPlugin = require("extract-text-webpack-plugin");
module.exports = {
    mode:'development',
    // 入口 js 文件，如果是一个字符串，就是要打包的路径；如果是对象，可以打包成多个模块
    entry:{
        first:'./src/first.js',
        second:'./src/second.js'
    },
    // 输出 js 文件
    output:{
        // 输出文件路径
        path:path.resolve(__dirname, './dist'),
        // 输出文件名称，当 entry 配置多个模块时需要使用 [name]，对应 entry 模块的 key
        filename:'[name].js',
        // 公共资源公共路径
        publicPath:'./'
    },
    // 定义对模块的处理规则
    module:{
        // 相当于旧版本的 loaders，配置编译规则
        rules:[
            //ES6 通过 babel 编译成 ES5
            {
                // 正则表达式，匹配编译的文件
                test:/\.(js|jsx)$/,
                // 排除的文件夹
                exclude:/node_modules/,
                use:[
                    {
                        loader:'babal-loader',
                        options:{
                            presets:[
                                ['env'], 'react', 'es2015', 'stage-0'],
                            plugins:['transform-runtime', 'add-module-exports'],
                        }
                    }
                ]
            },
```

```
            // 解析css，并把css添加到html的style标签
            {
                test:/.css$/,
                use:[
                    'style-loader', 'css-loader'
                ]
            },
            // 解析less，把less解析成浏览器可以识别的css语言
            {
                test:/.less$/,
                use:[
                    'style-loader', 'css-loader', 'less-loader'
                ]
            },
            // 解析图片资源
            {
                test:/.(jpg|png|gif|svg)$/,
                // 小于8MB处理成base64
                use:['url-loader?limit=8192&name=./[name].[ext]']
            },
            // 解析文件
            {
                test:/\.(gif|png|jpe?g|eot|woff|ttf|svg|pdf)$/,
                use:['file-loader'],
            },
        ]
    },
    // 定义插件，ExtractTextPlugin是将css独立引入变成link标签引入形式
    plugins:[new ExtractTextPlugin("style.css")],
    // 服务器配置
    devServer:{
        // 文件路径
        contentBase:path.resolve(__dirname, './dist'),
        // 主机地址
        host:'localhost',
        // 是否启动gzip压缩
        compress:true,
        // 端口号
        port:8000,
        // 是否热加载
        hot:true
```

            }
        };

以上是常用配置，更多配置请参考 webpack 官方文档。

7.2 页面跳转之 react-router-dom

虽然以上脚手架能够搭建出一个 React 示例，但在实际应用中，React 还需要配合路由模块来实现页面之间的跳转。

7.2.1 react-router-dom 路由配置

```
import React from 'react';
import {HashRouter, BrowserRouter,Route,Redirect,Switch,Link,NavLink} from 'react-router-dom';
import Home from './component/home';
import Detail from './component/detail';
const Router = () => (
<BrowserRouter>
<Route  exact  path="/"  component={Home}/>
    <Route  path="/detail"  name="detail" component={Detail}/>
    </BrowserRouter>
);
export default Router;
```

（1）路由配置属性主要有 path、name、component 等。

path：组件相对路径。

name：组件路径别名。

component：组件地址。

（2）在路由配置中，有两个属性 exact 和 strict 表示路径匹配。

exact 属性为 true 时匹配的路由有 "/detail/"，但是 URL 中的 "/detail" 匹配不到。

strict 属性为 true 时匹配的路由有 "/detail"，但是 "/detail/" 可以被匹配。

综上所述，要想严格匹配，就需要将 exact 和 strict 都设为 true。

7.2.2 react-router-dom 路由跳转方式

react-router-dom 路由跳转有两种方式。

（1）link 或 NavLink 方式，实质是个 a 标签，区别是后者可以切换时改变样式，用法如下。

```
<ul>
<li><NavLink exact to="/" activeStyle={{
    fontWeight:'bold',
        color:'red'
}}>home</NavLink>
</li>
<li><NavLink exact to="/detail" activeStyle={{
    fontWeight:'bold',
        color:'red'
}}>detail</NavLink>
</li>
</ul>
```

（2）直接使用 onClick 点击事件，用法如下。

```
<ul>
<li onClick={() => this.props.history.push({pathname:'detail'})}><div>home</div>
</li>
<li onClick={() => this.props.history.push({pathname:'home'})}><div>detail</div>
</li>
</ul>
```

7.2.3 react-router-dom 路由传参

上面提到两种路由跳转方式，相应的路由传参也有两种方式。

（1）组件跳转传参，使用 NavLink 进行配置，使用 to 属性进行跳转和传参，用法如下。

```
<ul>
<li><NavLink exact to="/" activeStyle={{
    fontWeight:'bold',
        color:'red'
}}>home</NavLink>
</li>
<li><NavLink exact to={{pathname:'detail',state:{id:2}}} activeStyle={{
    fontWeight:'bold',
        color:'red'
}}>detail</NavLink>
</li>
</ul>
```

(2)事件跳转传参,直接使用 state 作为 key,通过 this.props.history.location.state 获取参数,用法如下。

```
<ul>
<li onClick={() => this.props.history.push({pathname:'detail',state
:{id:1}})}><div>home</div>
</li>
<li onClick={() => this.props.history.push({pathname:'home',state:{
    id:0}})}><div>detail</div>
</li>
</ul>
```

7.3 前端组件库之 ant-design

React 框架主要是实现 UI 层,功能逻辑处理更多的是依赖第三方模块。而与 React 搭配得较为契合也较为稳定的第三方模块就是由蚂蚁金服前端团队开发的 ant-design 模块(移动端安装时模块名称是 antd-mobile)。

antd-mobile 库提供了丰富的组件,常用的有以下几个。

(1) Tabs 标签页,切换标签可定位到对应的内容,并将标签格式激活,效果如图 7-1 所示。

(2) Carousel 走马灯,主要应用于轮播图,效果如图 7-2 所示。

图 7-1

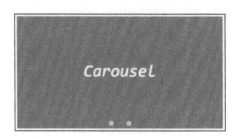

图 7-2

（3）DatePicker 日期选择，效果如图 7-3 所示。

（4）ListView 长列表，效果如图 7-4 所示。

图 7-3

图 7-4

以上列出来的 4 个组件只是比较常用的，整个 antd-mobile 中有几乎满足正常前端开发的所有场景的组件。更多组件的使用和介绍请参考官方文档。

7.4　CSS 预处理器之 Less

7.4.1　Less 特征

Less 是 CSS 预处理器，是对 CSS 的一种拓展。它具备动态语言的特点，如变量、运算、函数等，故也是一门动态语言。

7.4.2　Less 使用环境

Less 既可以在客户端（浏览器）使用，也可以在服务端（Node.js）使用。

客户端使用示例如下。

```
<link rel="stylesheet/less" href="style.less">
<script src="less.min.js"></script>
```

服务端使用示例如下。

```
npm install -g less
```

7.4.3　Less 语法

1. 变量

Less 中的变量都以符号 @ 开头。

如：@width : 100px; div{width : @width}。

2. 运算

Less 中可以对变量、颜色、函数等进行运算。

如：@width : 100px; div{width : @width/2}。

3. 字符串插值

Less 中可以将变量以类似 php 的语法 @{ } 插入字符串。

如：@url : 'img/index/' ; div{background : url（"@{url}search.png"）}。

4. 混合

在 Less 中可以将复用度较高的属性放在一起组成属性集并与其他属性混合使用。

如：.input{width : 100px;height : 30px};div>input{.input,color : red}。注意这里的 .input 可以理解为是一个变量，并不是 class。

5. 带参混合

可以在属性集中传入参数。

如：.input（@px）{width: @px;height: 30px}; div>input{.input（100px）,color: red}。

6. 命名空间

为了不与其他样式重名，Less 提供了命名空间，可将属性集放入命名空间内。

如：#public{.input（@px）{width:@px;height : 30px}; div>input{#public>.input（100px），color : red}}。

这里的 #public 就是一个命名空间，可以"#"或"."为符号开头。

7. 嵌套规则

嵌套规则指的是父子元素之间的样式可以进行嵌套，有利于维护。

如：body{background:white;div{color : red}}，div 作为 body 的子元素属性样式可以写在 body 内。

8. Color 函数

Less 提供了颜色运算函数，第一个参数为颜色值，第二个参数为颜色值的相对值。

如：lighten（red,10%）表示亮度浅 10%；

darken（red,10%）表示亮度深 10%；

saturate（red,10%）表示饱和度深 10%；

desaturate（red,10%）表示饱和度浅 10%；

fadein（red,10%）表示透明度深 10%；

fadeout（red,10%）表示透明度浅 10%；

fade（red,50%）表示透明度是第一个参数的 50%；

spin（red,10）表示颜色加深 10 度；

spin（red,-10）表示颜色减弱 10 度；

mix（red,blue）表示两个颜色的混合值。

示例：div{background:lighten（red,10%）}，得到的就是比 red 浅 10% 的颜色。

9. Math 函数

Less 提供了类似 JS 的 Math 函数用来处理数据类型的值。

如：round（2.5）表示四舍五入为 3；

ceil（2.4）表示向上取整为 3；

floor（2.5）表示向下取整为 2；

percentage（0.25）表示百分数 25%。

示例：div{border-radius:percentage（0.5）}，得到的就是 50% 的圆角。

10. 模式匹配

Less 中提供了类似 JS 的 switch 方法，可以用来做枚举。

如：.condition（@switch）{};.condition（red）{background:red}；.condition（blue）{background:blue}；

输入 .condition（red）得到 {background:red}。

11. 导引表达式

如：.condition（@px）when（@px>=300）{background: red}；.condition（@px）when（@px<300）{background:blue}；

输入 .condition（400）得到 {background:red}。

12. 作用域

Less 中与 JS 一样也有作用域的概念，所以会优先在局部作用域查找变量，如若没有，再往父级作用域查找。

如：@index : 100px;div{@index : 50px;width : @index}，这里的 div 的 width 就是 50px。

13. JavaScript 表达式

Less 也可以使用 JavaScript 表达式。

如：@var:`"hello".toUpperCase（）+ '!'`，输出 @var:"HELLO!"。

14. 注释

Less 中的注释与 CSS 一样，// 表示单行注释，/* */ 表示多行注释。

15. import

Less 中也可以用 import 引入模块。

如：@import "style.less"。

7.5 第三方模块安装

所有模块的安装都很简单，想要安装哪个模块只需要在 npm install 后面加上模块名即可。可执行 npm install react-router-dom antd 来下载路由模块与 antd 移动端模块，如图 7-5 所示。

```
C:\Users\caoxue\Desktop\secondary\secondary>npm install react-router-dom antd    同时下载路由与antd
npm WARN deprecated core-js@1.2.7: core-js@<2.6.8 is no longer maintained. Please, upgrade to core-js@
npm WARN @typescript-eslint/eslint-plugin@1.6.0 requires a peer of typescript@* but none is installed.
npm WARN @typescript-eslint/parser@1.6.0 requires a peer of typescript@* but none is installed. You mu
npm WARN @typescript-eslint/typescript-estree@1.6.0 requires a peer of typescript@* but none is instal
npm WARN ts-pnp@1.1.2 requires a peer of typescript@* but none is installed. You must install peer dep
npm WARN tsutils@3.14.1 requires a peer of typescript@>=2.8.0 || >= 3.2.0-dev || >= 3.3.0-dev || >= 3.
u must install peer dependencies yourself.
npm WARN optional SKIPPING OPTIONAL DEPENDENCY: fsevents@1.2.9 (node_modules\chokidar\node_modules\fse
npm WARN notsup SKIPPING OPTIONAL DEPENDENCY: Unsupported platform for fsevents@1.2.9: wanted {"os":"d
npm WARN optional SKIPPING OPTIONAL DEPENDENCY: fsevents@1.2.9 (node_modules\jest-haste-map\node_modul
npm WARN notsup SKIPPING OPTIONAL DEPENDENCY: Unsupported platform for fsevents@1.2.9: wanted {"os":"d
npm WARN optional SKIPPING OPTIONAL DEPENDENCY: fsevents@2.0.6 (node_modules\fsevents):
npm WARN notsup SKIPPING OPTIONAL DEPENDENCY: Unsupported platform for fsevents@2.0.6: wanted {"os":"d

+ react-router-dom@5.0.1
+ antd@3.20.7                 模块版本
added 99 packages from 159 contributors, updated 1 package and audited 904705 packages in 74.49s
found 0 vulnerabilities
```

图 7-5

第 8 章 前端开发前须知

【本章导读】
◎ 模块命名规则
◎ 选择器命名规则
◎ 公共样式提取
◎ 公共组件封装
◎ 界面分析

8.1 命名规则

8.1.1 模块命名规则

所有的命名均遵守语义化原则，即通过命名就知道其表达的意思。比如本项目划分为五大模块，分别为用户模块，即登录与注册（login 与 register）；商品模块，即首页（index）、发布商品（publish）、商品详情（indexDetail）等；支付模块、消息模块，以及个人中心模块。相应的模块文件和模块样式文件也可以用相同的名字，比如首页模块文件和首页模块样式文件可以命名为 index.js 和 index.css。

8.1.2 选择器命名规则

在样式文件中，会有各种各样的选择器元素，这里统一采用"模块-语义"命名规则，比如登录页面的一个登录按钮，选择器命名为"login-btn"。前面加模块的形式可以避免命名冲突，如果直接将登录按钮命名为"btn"，那么在其他页面也有按钮，就无法区别。采用"模块-语义"的形式，比如在注册页面也有注册按钮，就可以使用"register-btn"。这样既能避免命名冲突，又能区别模块，同时也遵守语义化原则。

8.2 公共样式提取

8.2.1 公共样式规则

公共样式命名规则为"cm-属性-属性值",如 cm-ptb-02。其中 cm 就表示 common,ptb 表示 padding-top 和 padding-bottom,02 则表示值为 0.2rem。

公共样式适用于较常见的样式,比如大部分页面左右边距相同,那么这个边距就可以提取为公共样式。

公共样式中内容结构大致分为初始化样式、整体布局、上、下、左、右边距、字体颜色和背景色、字体大小、图片大小、线条样式、圆角样式等。

初始化样式:不同的浏览器会对元素做一些细微的默认样式,所以为了考虑兼容性,在写项目样式前首先需要将部分元素的样式初始化。

整体布局:包含弹性布局(flex)、左右中布局(text-align),以及定位(position)等。

上、下、左、右边距:主要包括外边距(margin)以及内边距(padding)。

字体颜色和背景色:字体的颜色和元素的背景色。

字体大小:设置不同的字体大小。

图片大小:设置不同尺寸的图片大小。

线条样式:设置不同场景的线条样式。

圆角样式:设置不同场景的圆角样式。

8.2.2 公共样式文件

在 src 目录下新建 index.less 作为本项目的公共样式文件。

index.less 内容如下。

```
/* 初始化样式 */
@spacing:0.18rem;// 元素之间的核心间距
@theme-color:#cc0000;// 项目主题颜色
body{
    font:400 14px/1.5 Arial;// 初始化字体
}
* {
    box-sizing:border-box;
```

```
    -webkit-overflow-scrolling:touch;// 在 iOS 操作系统的手机中滑动更流畅
}
body, button, form, html, input, label,textarea{
    margin:0;
    padding:0;
}
//input 被填充时的样式
input:-webkit-autofill{
    -webkit-box-shadow:0 0 0 1000px #f5f5f9 inset;
}
//placeholder 的字体颜色
input::-webkit-input-placeholder,textarea::-webkit-input-placeholder{
    color:#999;
}
//Google Chrome 浏览器中去掉滚动条
::-webkit-scrollbar {
    display:none;
}
// 外边距
.cm-mt-009{
    margin-top:@spacing/2;
}
.cm-mt-018{
    margin-top:@spacing;
}
.cm-mt-036{
    margin-top:@spacing*2;
}
.cm-mt-080{
    margin-top:0.8rem;
}
.cm-ml-018{
    margin-left:@spacing;
}
.cm-mr-018{
    margin-right:@spacing;
}
.cm-mr-009{
    margin-right:@spacing/2;
}
```

```less
.cm-mtb-009{
   margin-top:@spacing/2;
   margin-bottom:@spacing/2;
}
.cm-mtb-018{
   margin:@spacing 0;
}
.cm-mlr-018{
   margin-left:@spacing;
   margin-right:@spacing;
}
.cm-m-018{
   margin:@spacing;
}
// 内边距
.cm-pt-018{
   padding-top:@spacing;
}
.cm-ptb-018{
   padding:@spacing 0;
}
.cm-ptb-009{
   padding-top:@spacing/2;
   padding-bottom:@spacing/2;
}
.cm-plr-009{
   padding-left:@spacing/2;
   padding-right:@spacing/2;
}
.cm-plr-018{
   padding:0 @spacing;
}
.cm-p-009{
   padding:@spacing/2;
}
.cm-p-018{
   padding:@spacing;
}
// 布局
.cm-flex{
   display:flex;
```

```css
}
.cm-flex-column{
  display:flex;
  flex-direction:column;
}
.cm-flex-1{
    flex:1;
}
.cm-flex-wrap{
  flex-wrap:wrap;
}
.cm-jc-c{
    justify-content:center;
}
.cm-jc-sa{
    justify-content:space-around;
}
.cm-jc-sb{
    justify-content:space-between;
}
.cm-jc-fe{
    justify-content:flex-end;
}
.cm-ai-c{
    align-items:center;
}
.cm-as-fe{
    align-self:flex-end;
}
.cm-as-fs{
    align-self:flex-start;
}
.cm-w-full{
  width:100%;
}
.cm-tx-l{
  text-align:left;
}
.cm-tx-r{
  text-align:right;
}
```

```less
.cm-tx-c{
  text-align:center;
}
// 字体颜色与背景色
.cm-c-main{
    color:@theme-color;
}
.cm-c-white{
    color:#fff;
}
.cm-c-333{
    color:#333;
}
.cm-c-666{
    color:#666;
}
.cm-c-999{
    color:#999;
}
.cm-bc-main{
    background:@theme-color;
}
.cm-bc-white{
    background:#fff;
}
.cm-bc-333{
    background:#333;
}
.cm-bc-666{
    background:#666;
}
.cm-bc-999{
    background:#999;
}
.cm-bc-ddd{
    background:#ddd;
}
// 字体大小
.cm-fs-020{
    font-size:0.20rem;
}
```

```css
.cm-fs-022{
    font-size:0.22rem;
}
.cm-fs-024{
    font-size:0.24rem;
}
.cm-fs-026{
    font-size:0.26rem;
}
.cm-fs-028{
    font-size:0.28rem;
}
.cm-fs-030{
    font-size:0.30rem;
}
.cm-fw-bold{
    font-weight:bold;
}
.cm-fs-height{
   line-height:0.4rem;
}
//图片尺寸
.cm-img-04{
    width:0.4rem;
    height:0.4rem;
}
.cm-img-05{
    width:0.5rem;
    height:0.5rem;
}
.cm-img-06{
    width:0.6rem;
    height:0.6rem;
}
.cm-img-08{
    width:0.8rem;
    height:0.8rem;
}
.cm-img-10{
    width:1rem;
    height:1rem;
```

```css
}
.cm-img-12{
    width:1.2rem;
    height:1.2rem;
}
.cm-img-14{
    width:1.4rem;
    height:1.4rem;
}
.cm-img-16{
    width:1.6rem;
    height:1.6rem;
}
.cm-img-banner{
    width:100%;
    height:4rem;
}
// 边框样式
.cm-border-ddd {
  border:1px solid #ddd;
}
.cm-border-bottom-ddd {
  border-bottom:1px solid #ddd;
}
.cm-border-top-eee {
  border-top:1px solid #eee;
}
.cm-border-bottom-eee {
    border-bottom:1px solid #eee;
}
// 圆角样式
.cm-border-radius-10{
    border-radius:0.1rem;
}
.cm-border-radius-half{
    border-radius:50%;
}

.cm-space-line{
    height:0.18rem;
    background:#eee;
```

```less
}
// 按钮样式
.cm-btn-main{
    background:@theme-color;
    padding:0.05rem 0.2rem;
    color:#fff;
    border-radius:0.1rem;
}
.cm-btn-main-higher{
    background:@theme-color;
    padding:0.05rem 0.2rem;
    color:#fff;
    border-radius:0.1rem;
}
.cm-btn-border-333{
    padding:0.05rem 0.2rem;
    border:1px solid #333;
    color:#333;
    border-radius:0.1rem;
}
.cm-btn-border-main{
    border:1px solid @theme-color;
    padding:0.05rem 0.2rem;
    color:@theme-color;
    border-radius:0.1rem;
}
.cm-btn-border-999{
    border:1px solid #999;
    padding:0.05rem 0.2rem;
    color:#999;
    border-radius:0.1rem;
}
// 底部悬浮定位
.cm-bottom-position{
    position:fixed;
    bottom:0;
    left:0;
    right:0;
    z-index:999;
    width:100%;
    display:flex;
```

```
align-items:center;
background:#fff;
padding:0.18rem;
}
```

8.3 公共组件封装

在使用 React 框架进行开发时，虽然同时使用 ant-design 组件库能基本满足业务场景需求。但是作为一个开发者，不能永远重复使用别人的"轮子"，自己也要有创造轮子的本领才行。所以在实际开发中，有一些共同的样式、布局或者交互效果都可以封装成公共组件。

在 src 目录下新建 share 文件夹，用来存放所有公共组件及资源文件，分别添加以下 9 个公共组件和样式。在 src/share 目录下新建 images 文件夹，将公共组件的图片文件存放进去。在 src/share 目录下新建 index.js，将所有公共组件导出。在 src/share 目录下新建 default.less 文件，作为样式的公共变量。

为了使公共组件也可以运用到其他项目，公共组件的单位采用 px，所有的公共组件样式都使用 Less。

index.js 内容如下。

```
export { default as Select } from './select/index';
export { default as Title } from './title/index';
export { default as List } from './list/index';
export { default as Input } from './input/index';
export { default as Search } from './search/index';
export { default as Type } from './type/index';
export { default as Button } from './button/index';
export { default as TabBottom } from './tab-bottom/index';
export { default as Card } from './card/index';
```

default.less 内容如下。

```
@hd:1px; // 基本单位

// 字体颜色
```

```less
@c-main:#cc0000;
@c-gray:#ddd;
@c-white:#fff;
@c-ccc:#ccc;
@c-eee:#eee;
@c-999:#999;
@c-333:#333;

// 背景色
@bc-main:#cc0000;
@bc-black:#333;
@bc-gray:#ddd;
@bc-body:#f5f5f9;
@bc-white:#fff;

// 图片尺寸
@size-10:10 * @hd;
@size-12:12 * @hd;
@size-14:14 * @hd;
@size-16:16 * @hd;
@size-20:20 * @hd;
@size-30:30 * @hd;
@size-40:40 * @hd;
@size-50:50 * @hd;

// 边框色
@border-c-base:#ddd;
```

下面介绍本项目中封装的共同组件。

8.3.1　Button 组件

Button 组件有两种样式,第一种是 primary,红字红框白底,宽度为屏幕的 40%;第二种是 fill,白字红底,宽度占满屏幕(除去边距)。

组件说明:

实现两种 button 按钮,一种实心铺满宽度为 100%,另一种空心宽度为 30%(边框)。

效果展示:

实心铺满宽度为 100% 的效果如图 8-1 所示。

空心宽度为 30%（边框）的效果如图 8-2 所示。

图 8-1

图 8-2

API 属性说明如表 8-1 所示。

表 8-1 属性说明

属性	说明	类型	默认值
type	按钮类型，可选值为 primary / fill 或者不设	string	primary
onClick	点击事件	点击按钮的回调函数	(e: Object): void
className	样式类名	string	无

index.js 内容如下：

```
import React from 'react';
import classnames from 'classnames';
import './index.css';
import {defaultProps} from './defaultProps';
import TouchFeedback from 'rmc-feedback';
// 两个中文字符的正则表达式
const rxTwoCNChar = /^[\u4e00-\u9fa5]{2}$/;
// 判断是否是两个字符
const isTwoCNChar = rxTwoCNChar.test.bind(rxTwoCNChar);
// 判断是否是字符类型
function isString(str) {
    return typeof str === 'string';
}
// 判断如果是两个中文字符则插入空格
function insertSpace(child) {
    if (isString(child.type) && isTwoCNChar(child.props.children)) {
        return React.cloneElement(
            child,
            {},
            child.props.children.split('').join(' '),
        );
    }
    if (isString(child)) {
```

```javascript
        if (isTwoCNChar(child)) {
            child = child.split('').join(' ');
        }
    }
    return child;
}
export default class Button extends React.Component{
    static defaultProps = defaultProps;
    handleClick(){
        const { onClick } = this.props;
        if (onClick) {
            onClick();
        }
    };
    render() {
        const {prefixCls,type,children,className} =this.props;
        // 通过classnames方法将所有的class整合
        const wrapCls = classnames(prefixCls, className, {
            ['${prefixCls}-btn']:(type === 'primary'||!type),
            ['${prefixCls}-btn-fill']:type === 'fill',
            ['${prefixCls}-btn-half']:type === 'half',
        });
        // 调用insertSpace方法判断，如果是两个中文字符则插入空格
        const kids = React.Children.map(children, insertSpace);
        return(
            <TouchFeedback
                activeClassName={prefixCls+"-active"}
            ><div className={wrapCls} onClick={()=>this.handleClick()}
>{kids}</div></TouchFeedback>
        )
    }
}

function noop() {}
export const defaultProps = {
    prefixCls:'s-button',
    onClick:noop,
    type:"",
    children:[],
    className:""
};
```

```
defaultProps.js
function noop() {}
export const defaultProps = {
    prefixCls:'s-button',
    onClick:noop,
    type:"",
    children:[],
    className:""
};
```

index.less 内容如下:

```less
@import '../default';
@prefixCls:s-button;
.@{prefixCls} {
  display:flex;
  justify-content:center;
  &-btn{
    text-align:center;
    height:(@size-40)-5;
    line-height:(@size-40)-5;
    border:1px solid @c-main;
    border-radius:@size-10;
    color:@c-main;
    width:30%;
    margin-left:auto;
    margin-right:auto;
  }
  &-active {
    background-color:#eee;
  }
}
.@{prefixCls}-btn-fill{
  height:@size-40;
  line-height:@size-40;
  background:@c-main;
  color:@c-white;
  border-radius:@size-10;
  text-align:center;
  font-size:@size-16;
  font-weight:bold;
```

```
}
.@{prefixCls}-btn-half{
  height:@size-40;
  line-height:@size-40;
  background:@c-main;
  color:@c-white;
  border-radius:@size-10;
  text-align:center;
  font-size:@size-16;
  font-weight:bold;
  width:48%;
}
```

8.3.2　Card 组件

Card 组件主要是将列表布局成卡片形式。

组件说明：

使列表成为卡片形式的布局。

效果如图 8-3 所示。

图 8-3

API 属性说明如表 8-2 所示。

表 8-2　属性说明

属性	说明	类型	默认值
onClick	点击卡片的回调	Function	无

index.js 内容如下：

```js
import React from 'react';
import {defaultProps} from './defaultProps';
import './index.css';
export default class Card extends React.Component {
    static defaultProps = defaultProps;
    render(){
        let {onClick,prefixCls} = this.props;
        return(
            <div className={prefixCls} onClick={()=>onClick?onClick():undefined}>
                {this.props.children}
            </div>
        )
    }
}
```

```js
defaultProps.js
function noop() {}
export const defaultProps = {
    prefixCls:'s-card',
    onClick:noop
};
```

index.less 内容如下：

```less
@import '../default';
@prefixCls:s-card;

.@{prefixCls}{
  //display:flex;
  margin:(@size-10)-2 0;
  padding:(@size-10)/(@size-10);
  box-shadow:2px 2px 2px 2px #eee;
  background:#fff;
```

```
    border-radius:5px;
}
```

8.3.3 Input 组件

Input 组件主要是重写项目中原生 input 文本框的默认样式。

组件说明:

实现 Input 文本框。

效果展示:

左边是图标,右边是文本框及提示,如图 8-4 所示。

API 属性说明如表 8-3 所示。

图 8-4

表 8-3 属性说明

属性	说明	类型	默认值
type	文本框类型	any	text
src	左边的图标	string	无
placeholder	文本框提示	string	无
value	文本框值	any	无
defaultValue	文本框默认值	any	无
onChange	文本框值改变时的事件	Function	空函数
maxLength	文本框输入值长度	Number	11
style	文本框样式	Object	无
className	文本框样式类名	string	无
renderRight	右边渲染的元素	HTML 元素	null

index.js 内容如下:

```
import React from 'react';
import classnames from 'classnames';
import './index.css';
import {defaultProps} from './defaultProps';
export default class Input extends React.Component {
    static defaultProps = defaultProps;
    change(event){
        const {onChange}=this.props;
```

```
            onChange(event.target.value,event);
        }
        render() {
            const {prefixCls,label,defaultValue,placeholder,type,src,renderRight,className,maxLength,style,disabled} =this.props;
            return (
                <div className={prefixCls}>
                    <div className={prefixCls+"-wrap"}>
                    {src? <img src={src} alt="" className={prefixCls+"-icon"}/>:label?<label htmlFor="" className={prefixCls+"-label"}>{label}</label>:null}
                        <input type={type?type:"text"}
                            disabled={disabled}
                            onChange={(e)=>this.change(e)}
                            className={classnames(prefixCls+"-input",className)}
                            placeholder={placeholder}
                            maxLength = {maxLength}
                            style={style}
                            defaultValue={defaultValue}
                        />
                    </div>
                    {renderRight?renderRight:null}
                </div>
            )
        }
    }
```

defaultProps.js 内容如下:

```
function noop() {}
export const defaultProps = {
    prefixCls:'s-input',
    onClick:noop,
    label:"",
    placeholder:"",
    type:"",
    src:"",
    renderRight:null,
    className:"",
```

```
    maxLength:11,
    style:{}
};
```

index.less 内容如下:

```
@import '../default';
@prefixCls:s-input;

/* 默认搜索 bar */
.@{prefixCls} {
  display:flex;
  justify-content:space-between;
  .@{prefixCls}-wrap{
    display:flex;
    align-items:center;
    width:100%;
    flex:1;
    .@{prefixCls}-label{
      margin-right:@size-10;
    }
    .@{prefixCls}-icon{
      width:@size-20;
      margin-right:@size-10;
    }
    .@{prefixCls}-input{
      flex:1;
      border:none;
      height:@size-20+5;
      font-size:@size-14;
      background:@bc-body;
      text-align:left;
    }
  }
}
```

8.3.4 List 组件

List 组件主要是封装行样式。

组件说明:

List 列表主要实现底部下划线和左右留空。

图 8-5

效果展示:

List 组件通常与其他组件结合使用,图 8-5 所示是 List+Input 和 List+Select 的使用效果。

API 属性说明如表 8-4 所示。

表 8-4　属性说明

属性	说明	类型	默认值
onClick	某一行的点击事件	Function	空函数

index.js 内容如下:

```
import React from 'react';
import Item from './ListItem';
import {defaultProps} from './defaultProps';
import './index.css';
export default class List extends React.Component {
    static defaultProps = defaultProps;
    render(){
        let {onClick,prefixCls,border} = this.props;
        return(
            <div className={border?prefixCls:prefixCls+'-no-border'} onClick={(e)=>onClick(e)}>
                {this.props.children}
            </div>
        )
    }
}
List.Item = Item;
```

defaultProps.js 内容如下:

```
function noop() {}
export const defaultProps = {
    prefixCls:'s-list',
    border:true,
```

```
    onClick:noop,
};
```

index.less 内容如下：

```less
@import '../default';
@prefixCls:s-list;

.@{prefixCls} {
  padding:@size-10 0;
  border-bottom:1px solid @border-c-base;
  .@{prefixCls}-item{
    display:flex;
    justify-content:space-between;
    width:100%;
    .@{prefixCls}-item-left{
      display:flex;
      align-items:center;
      .@{prefixCls}-item-left-icon{
        width:@size-20;
        height:@size-20;
        margin-right:(@size-10)-5;
      }
      .@{prefixCls}-item-left-text{
        color:@c-333;
      }
    }
    .@{prefixCls}-item-right{
      display:flex;
      align-items:center;
      margin-left:@size-10;
      color:@c-999;
      .@{prefixCls}-item-right-icon{
        width:@size-20;
        height:@size-20;
      }
      .@{prefixCls}-item-right-text{
        color:@c-ccc;
      }
    }
  }
}
```

```less
  }
  /* 默认搜索 bar */
  .@{prefixCls}-no-border {
    //display:flex;
    padding:@size-10 0;
    .@{prefixCls}-item{
      display:flex;
      justify-content:space-between;
      width:100%;
      .@{prefixCls}-item-left{
        display:flex;
        align-items:center;
        .@{prefixCls}-item-title{
          color:@c-333;
        }
      }
      .@{prefixCls}-item-right{
        display:flex;
        align-items:center;
        color:@c-999;
      }
      .@{prefixCls}-item-left-icon{
        width:@size-20;
        height:@size-20;
        margin-right:(@size-10)-5;
      }
      .@{prefixCls}-item-right-icon{
        width:@size-20;
        height:@size-20;
      }
    }
  }
```

ListItem.js 内容如下:

```js
import React from 'react';
import right from '../images/list/right.png';
import {defaultProps} from './defaultProps';
export default class ListItem extends React.Component {
    static defaultProps = defaultProps;
    onRightClick(){
```

```
                const {onRightClick} =this.props;
                if (onRightClick) {
                    onRightClick();
                }
            }
            render(){
                const {prefixCls,leftIcon,leftTitle,rightTitle,rightIcon,is
NotRightIcon} = this.props;
                return(
                    <div className={prefixCls+"-item"}>
                        <div className={prefixCls+"-item-left"}>
                            {leftIcon?<img src={leftIcon} alt="" className=
{prefixCls+"-item-left-icon"}/>:null}
                            <span className={prefixCls+"-item-left-
text"}>{leftTitle}</span>
                        </div>
                        <div className={prefixCls+"-item-
right"} onClick={()=>this.onRightClick()}><span className={prefixCls+"-
item-right-text"}>{rightTitle}</span>
                            {isNotRightIcon?null:
                                rightIcon?<img src={rightIcon} alt="" className=
{prefixCls+"-item-right-
icon"}/>:<img src={right} alt=""  className={prefixCls+"-item-right-
icon"}/>
                            }
                        </div>
                    </div>
                )
            }
        }
```

8.3.5 Search 组件

Search 组件主要是封装带有搜索功能的文本框。

组件说明：

实现文本框搜索功能。

效果展示：

输入关键字，可以通过键盘的搜索按钮完成搜索功能，如图 8-6 所示。

图 8-6

API 属性说明如表 8-5 所示。

表 8-5 属性说明

属性	说明	类型	默认值
searchIcon	搜索图标	string	不传为默认图标
onSearch	点击搜索图标触发事件	string	(e: Object): void
onRightClick	点击右边图标触发事件	string	(e: Object): void
rightIcon	右边的图标	string	不传为默认图标
closeIcon	关闭的图标	string	不传为默认图标
placeholder	文本框提示文字	string	空

index.js 内容如下：

```
import React from 'react';
import './index.css';
import {defaultProps} from './defaultProps';
export default class Search extends React.Component {
    static defaultProps = defaultProps;
    constructor(props) {
        super(props);
        this.state = {
            isShowClose:false,// 是否显示文本框右边的关闭图标
        }
    }
    // 点击左边搜索图标触发的事件
    leftClick() {
        this.callBack();
    }
    // 文本框回调
    callBack(){
        const {onSearch} =this.props;
        var key = this.refs.input.value;
        if (onSearch) {
            onSearch(key);
        }
    }
    // 执行右边图标的事件
    rightClick() {
        const {onRightClick} =this.props;
        if (onRightClick) {
```

```
            onRightClick();
        }
    }
    // 点击文本框右边的关闭图标时，清除搜索内容并删除图标
    closeValue() {
        this.refs.input.value = '';
        this.setState({
            isShowClose:false
        });
    }
    // 点击文本框进行关键字搜索时，右边出现关闭图标
    onChange(event) {
        if (event.target.value === '') {
            this.setState({
                isShowClose:false
            })
        } else {
            this.setState({
                isShowClose:true
            })
        }
    }
    // 按下键盘时触发的事件
    keyDown(e){
        if(e.keyCode === 13){
            e.preventDefault();
            this.refs.input.blur();
            this.callBack();
        }
    }
    render() {
        const {
            prefixCls,
            searchIcon,
            rightIcon,
            placeholder,
            closeIcon
        } =this.props;
        return (
            <div className={prefixCls}>
                <div className={prefixCls+"-layout"}>
```

```jsx
                        <div className={prefixCls+"-content-left"}>
                            <img src={searchIcon} alt="" className=
{prefixCls+"-search-icon"}
                                 onClick={() => this.leftClick()}/>
                            <form action="#" className={prefixCls+"-form"}>
                            <input type="search" ref="input"  className=
{prefixCls+"-input"}
                                   placeholder={placeholder}
                                   onChange={(e) => this.onChange(e)}
                                   onKeyDown={(e)=>this.keyDown(e)}
                            />
                            </form>
                            {this.state.isShowClose ?
                                <img src={closeIcon} alt="" className=
{prefixCls+"-close-icon"}
                                     onClick={() => this.
closeValue()}/> :null
                            }
                        </div>
                        {rightIcon?<img src={rightIcon} alt="" className=
{prefixCls+"-right-icon"} onClick={() => this.rightClick()}/>:null}
                    </div>
                </div>
            )
        }
    }
```

defaultProps.js 内容如下:

```jsx
import search from '../images/search/search.png';
import close from '../images/search/close.png';
function noop() {}
export const defaultProps = {
    prefixCls:'s-search',
    placeholder:'',
    leftClick:noop,
    rightClick:noop,
    searchIcon:search,
    closeIcon:close,
    rightIcon:"",
};
```

index.less 内容如下：

```less
@import '../default';
@prefixCls:s-search;

.@{prefixCls} {
  height:@size-40+5;
  .@{prefixCls}-layout {
    display:flex;
    align-items:center;
    position:fixed;
    left:0;
    background:@bc-main;
    right:0;
    padding:0  10px;
    z-index:99;
    height:@size-40+5;
    .@{prefixCls}-content-left{
      display:flex;
      width:100%;
      .@{prefixCls}-search-icon{
        width:@size-14;
        height:@size-14;
        position:absolute;
        top:50%;
        transform:translate(50%, -50%);
      }
      .@{prefixCls}-form{
        width:100%;
      }
      .@{prefixCls}-input{
        background:@bc-white;
        border:1px solid #ddd;
        border-radius:@size-20;
        height:@size-30;
        padding:@size-10 @size-10 @size-10 @size-20+5;
        width:100%;
        margin-right:@size-10;
        font-size:@size-14;
      }
      .@{prefixCls}-input::-webkit-input-placeholder{
```

```
      color:@c-ccc;
    }
    .@{prefixCls}-close-icon{
      width:@size-14;
      height:@size-14;
      position:absolute;
      top:50%;
      right:@size-40+5;
      transform:translate(-50%, -50%);
    }
  }
  .@{prefixCls}-right-icon{
    margin-left:@size-10;
    width:@size-20;
    height:@size-20;
  }
 }
}
```

8.3.6 Select 组件

Select 组件主要是将 select 选项使用下拉列表框的形式展示。

组件说明：

选择选项时出现下拉列表框。

效果展示：

点击选项时出现下拉列表框，如图 8-7 所示。

API 属性说明如表 8-6 所示。

图 8-7

表 8-6 属性说明

属性	说明	类型	默认值
options	选项数组	Array	[]
babel	选项标题	string	无
onChange	选中某一行的点击事件	void	无

index.js 内容如下：

```
import React from 'react';
```

```js
import right from '../images/select/right.png';
import select from '../images/select/select.png';
import './index.css';
import close from '../images/select/close.png';
import {defaultProps} from './defaultProps';
export default class Select extends React.Component {
    static defaultProps = defaultProps;
    constructor(props) {
        super(props);
        this.state = {
            isShow:false,// 是否显示下列拉列框,初始隐藏
            active:(props.activeOne && props.activeOne.type-1)||0,
// 默认选中某一行
            value:'',// 类别名称
            id:'',// 类别 id
            bottomHeight:250,// 下列拉列框高度
        }
    }
    componentWillReceiveProps({options,onChange}){
        const isContainedInNumbers = options.some(number => number.id === this.state.id);
        if (isContainedInNumbers) {
            return;
        }
        this.state.value = this.props.activeOne && this.props.activeOne.name||options[0].name;// 默认显示第一个类别
        this.state.id = this.props.activeOne && this.props.activeOne.type||options[0].id;// 默认显示第一个类别 id
        // 由于更新父组件 state,导致父组件重新 render,// 而 React 在 render 时,不管子组件属性是否改变,都会调用 componentWillReceiveProps,// 因此会导致 componentWillReceiveProps 死循环。// 不过没关系,前面加上了终止条件
        onChange(options[0]);
    }
    // 显示下拉列表框
    showBox(str) {
        this.setState({
            isShow:!this.state.isShow
        },()=>{
            if(str === "setHeight"){
                this.setHeight(this)
            }
```

```jsx
        })
    }
    // 设置下拉列表框的高度
    setHeight(that){
        var height = this.refs && this.refs.box.offsetHeight;
        console.log(height);
        if(height>=this.state.bottomHeight){
            console.log(that.refs.box);
            that.refs.box.style.height = this.state.bottomHeight+"px";
            that.refs.box.style.overflow = "scroll";
        }
    }
    // 关闭下拉列表框
    closeBox(){
        this.setState({
            isShow:false
        })
    }
    // 选择了某一行触发的事件
    change(item,index){
        const {onChange}=this.props;
        onChange(item);
        this.setState({
            value:item.name,
            active:index
        })
    }
    render() {
        const {prefixCls,options, babel} =this.props;
        return (
            <div className={prefixCls}>
                <div>{babel}</div>
                <div className={prefixCls+'-wrap'} onClick={() => this.showBox("setHeight")}>
                    <div>{this.state.value}</div>
                    <img src={right} alt="" className={prefixCls+'-arrow'}/>
                </div>
                {this.state.isShow ?
                    <div className={prefixCls+'-modal'}>
                        <div className={prefixCls+'-content'}
```

```jsx
ref="box" onClick={()=>this.showBox()}>
                                <div className={prefixCls+'-height'}>
                            <div className={prefixCls+'-item'}>
                                <div className={prefixCls+'-item-left'}>
                                    <img className={prefixCls+'-item-icon'} src={select} alt="" onClick={()=>this.closeBox()}/>
                                    <span className="cm-ml-10 cm-c-main">请选择分类</span>
                                </div>
                                <img className={prefixCls+'-item-icon'} src={close} alt="" onClick={()=>this.closeBox()}/></div>
                                {options.map((item,index) => {
                                    return <div key={index} className={this.state.active === index? prefixCls+'-label-active':prefixCls+'-label'} onClick={()=>this.change(item,index)}>{item.name}</div>
                                })}
                                </div></div> :null}
            </div>
        )
    }
}

/*defaultProps.js*/
export const defaultProps = {
    prefixCls:'s-select',
    options:[],
    babel:""
};

/*index.less*/
@import '../default';
@prefixCls:s-select;

.@{prefixCls} {
  display:flex;
  align-items:center;
  justify-content:space-between;
  width:100%;
  .@{prefixCls}-wrap{
      display:flex;
```

```less
      align-items:center;
      .@{prefixCls}-arrow{
        width:@size-20
      }
    }
  }
  .@{prefixCls}-modal{
    position:absolute;
    left:0;
    right:0;
    top:0;
    bottom:0;
    z-index:9999;
    text-align:center;
    background:rgba(50,50,50,0.5);
    .@{prefixCls}-content{
      position:absolute;
      left:0;
      right:0;
      bottom:0;
      z-index:9999;
      text-align:center;
      background:#ffffff;
      .@{prefixCls}-height{
        height:@size-40;
      }
      .@{prefixCls}-item{
        color:@c-main;
        display:flex;
        align-items:center;
        justify-content:space-between;
        position:fixed;
        left:0;
        right:0;
        background:#ffffff;
        height:@size-40;
        line-height:@size-40;
        .@{prefixCls}-item-left{
          display:flex;
          align-items:center;
          margin-left:@size-10;
        }
      }
```

```less
      .@{prefixCls}-item-icon{
        width:@size-20;
        margin-right:@size-10;
      }
    }
    .@{prefixCls}-label{
      height:@size-40;
      line-height:@size-40;
      color:@c-999;
      border-top:1px solid @border-c-base;
    }
    .@{prefixCls}-label-active{
      height:@size-40;
      line-height:@size-40;
      color:@c-main;
      border-top:1px solid @border-c-base;
    }
  }
}
```

8.3.7 TabBottom 组件

TabBottom 组件主要用来封装底部 tab 切换按钮。

组件说明：

实现底部 tab 切换。

效果展示：

底部 tab 切换，改变激活样式，切换到对应的页面，如图 8-8 所示。

图 8-8

API 属性说明如表 8-7 和表 8-8 所示。

表 8-7 属性说明

属性	说明	类型	默认值
history	将父组件的 history 传递到子组件	Object	无
defaultTab	默认选中的 tab	string	1
onClick	点击 tab 触发的事件	(e: Object): void	空函数
tabList	数组，将底部的 tab 名称、路由等信息组合到 tabList	Array	[]

表 8-8 属性说明

属性	说明	类型	默认值
title	底部的 tab 文字	string	""
path	点击 tab 之后跳转到对应的页面路径名称	string	""
icon	未选中时的图标	string	""
checkedIcon	选中时的图标	string	""

index.js 内容如下：

```
import React from 'react';
import myAct from '../images/tab-bottom/my.png';
import my from '../images/tab-bottom/my-gray.png';
import msgAct from '../images/tab-bottom/message.png';
import msg from '../images/tab-bottom/message-gray.png';
import index from '../images/tab-bottom/index-gray.png';
import indexAct from '../images/tab-bottom/index.png';
import publish from '../images/tab-bottom/pubilsh-gray.png';
import publishAct from '../images/tab-bottom/publish.png';
import {defaultProps} from './defaultProps';
import './index.css';
export default class TabBottom extends React.Component {
    static defaultProps = defaultProps;
    constructor(props) {
        super(props);
        this.tabList = props.tabList.length>0?props.tabList:[
            {label:"首页",icon:index,iconAct:indexAct,path:"index"},
            {label:"发布",icon:publish,iconAct:publishAct,path:"publish"},
            {label:"消息",icon:msg,iconAct:msgAct,path:"message"},
            {label:"我的",icon:my,iconAct:myAct,path:"personal"},
        ];
        this.state = {
            activeKey:props.activeNum//默认显示第一个 tab
        }
    }
    //点击 tab 时触发事件
    changeTab(path){
        this.props.history.push(path);
    }
    render() {
```

```
            const {prefixCls} = this.props;
            return (
                <div className={prefixCls}>
                    <div className={prefixCls+"-position-fixed"}>
                        {this.tabList.map((item,index)=>{
                            return(<div key={index} className={index 
=== this.state.activeKey?prefixCls+"-tab-active":prefixCls+"-tab"} onClick=
{()=>this.changeTab(item.path)}>
                                <img className={prefixCls+"-icon"} 
src={index === this.state.activeKey?item.iconAct:item.icon} alt=""/>
                                <span className={prefixCls+"-span"}>{item.
label}</span>
                            </div>)
                        })}
                    </div>
                </div>
            )
        }
    }
```

defaultProps.js 内容如下：

```
export const defaultProps = {
    prefixCls:'s-tab-bottom',
    tabList:[],
    activeNum:0
};
```

index.less 内容如下：

```
@import '../default';
@prefixCls:s-tab-bottom;

.@{prefixCls} {
  height:@size-50;
  .@{prefixCls}-position-fixed{
    display:flex;
    height:@size-50;
    justify-content:space-around;
    position:fixed;
```

```less
        bottom:0;
        left:0;
        right:0;
        background:#ffffff;
        z-index:1000;
    }
    .@{prefixCls}-tab {
        color:#999;
        font-size:@size-14;
        display:flex;
        flex-direction:column;
        justify-content:center;
        align-items:center;
    }
    .@{prefixCls}-tab-active {
        color:#cc0000;
        font-size:@size-14;
        display:flex;
        flex-direction:column;
        justify-content:center;
        align-items:center;
    }
    .@{prefixCls}-icon{
        width:(@size-20)+2;
    }
    .@{prefixCls}-span{
        font-size:@size-12;
        margin-top:3px;
    }
}
```

8.3.8　Title 组件

Title 组件主要用来封装标题样式。

组件说明：

实现两种 title 标题，一种有主题填充色，没有返回图标，适用于一级页面的标题；另一种没有主题填充色，有返回图标，适用于非一级页面的标题。

效果展示：

有主题填充色，一级页面标题效果，如图 8-9 所示。

无主题填充色，非一级页面标题效果，如图 8-10 所示。

图 8-9

图 8-10

API 属性说明如表 8-9 所示。

表 8-9　属性说明

属性	说明	类型	默认值
title	标题文字	String	无
isHome	是否是一级页面	Boolean	false
onLeftClick	点击返回图标触发事件	点击返回图标的回调函数	(e: Object): void
style	样式类名	string	无

index.js 内容如下：

```
import React from 'react';
import './index.css';
import backBlock from '../images/title/back-block.png';
import {defaultProps} from './defaultProps';
export default class Title extends React.Component {
    static defaultProps = defaultProps;
    // 点击返回图标触发事件
    onLeftClick() {
        const {onLeftClick,that} =this.props;
        if (onLeftClick) {
            onLeftClick();
        }else {
            that.props.history.goBack()
        }
    }
    onRightClick(){
        const {onRightClick} =this.props;
        if (onRightClick) {
            onRightClick();
        }
    }
    render() {
        const {title,prefixCls,isHome,style,rightTitle,rightIcon} =this.props;
```

```jsx
            if (isHome) {
                return (
                    <div className={prefixCls}>
                        <div className={prefixCls+"-warp-home"}>{title}</div>
                    </div>
                )
            } else {
                return (
                    <div className={prefixCls}>
                        <div className={prefixCls+"-warp"} style={style}>
                            <div className={prefixCls+"-left"}>
                                <img src={backBlock} alt="" className={prefixCls+"-left-icon"} onClick={() => this.onLeftClick()}/>
                            </div>
                            <span>{title}</span>
                            <div className={prefixCls+"-right"} onClick={()=>this.onRightClick()}>{rightTitle?rightTitle:(rightIcon?
                                <img src={rightIcon} className={prefixCls+"-right-icon"} alt=""/>:'')}</div>
                        </div>
                    </div>
                )
            }
        }
    }
```

defaultProps.js 内容如下：

```jsx
export const defaultProps = {
    prefixCls:'s-title',
    title:'',
    isHome:false,
};
```

index.less 内容如下：

```less
@import '../default';
@prefixCls:s-title;
.@{prefixCls} {
  height:@size-40;
```

```less
.@{prefixCls}-warp-home {
  position:fixed;
  background:@bc-main;
  color:@bc-white;
  left:0;
  right:0;
  height:@size-40;
  line-height:@size-40;
  text-align:center;
  font-size:@size-16;
  font-weight:bold;
  z-index:999;
}
.@{prefixCls}-left,.@{prefixCls}-right{
  width:@size-50;
}
.@{prefixCls}-right{
  text-align:right;
  font-size:@size-14;
}
.@{prefixCls}-left-icon,.@{prefixCls}-right-icon{
  width:@size-20;
  vertical-align:middle;
}
.@{prefixCls}-warp {
  display:flex;
  background:#f5f5f9;
  justify-content:space-between;
  align-items:center;
  position:fixed;
  left:0;
  right:0;
  height:@size-40;
  line-height:@size-40;
  padding:0 (@size-10)-1;
  font-size:@size-16;
  z-index:999;
  color:#333;
}
}
```

8.3.9 Type 组件

Type 组件主要用来封装分类选择。

组件说明：

实现分类选择。

效果展示：

点击分类类别，获取不同的分类列表，分类标签将位移到屏幕中心区域，如图 8-11 所示。

图 8-11

API 属性说明如表 8-10 所示。

表 8-10 属性说明

属性	说明	类型	默认值
typeList	类别数组	Array	[]
onTypeClick	点击类型的回调事件	Function	(e: Object): void

index.js 内容如下：

```
import React from 'react';
import './index.css';
import {defaultProps} from './defaultProps';
export default class Type extends React.Component {
    static defaultProps = defaultProps;
    constructor(props) {
        super(props);
        this.scrollLeft = 0;
        this.state = {
            activeType:0// 默认选中第一个类型
        }
    }
    // 点击类型时触发事件
    onTypeClick(index,item,e) {
        var el = e.target.getBoundingClientRect();
        this.setState({
            activeType:index
        });
        var count = 0;
        var scroll = this.refs.scroll;
        var slideHalf = window.innerWidth/2;
        // 滑动时同时位移到屏幕中心区域
```

```jsx
            if(el.x>slideHalf){
                count = 0;
                var timer = setInterval(()=>{
                    if(count <= (el.x-slideHalf)){
                        scroll.scrollTo(this.scrollLeft + count,0);
                        count+=1;
                    } else {
                        clearInterval(timer)
                    }
                },10);
            }else {
                count = 0;
                var timer1 = setInterval(()=>{
                    if(count <= Math.abs(el.x-slideHalf)){
                        scroll.scrollTo(this.scrollLeft - count,0);
                        count+=1;
                    }else {
                        clearInterval(timer1)
                    }
                },10);
            }
            this.scrollLeft = scroll.scrollLeft;
            const {onTypeClick} =this.props;
            if (onTypeClick) {
                onTypeClick(item);
            }
        }
        render() {
            const {prefixCls,typeList} =this.props;
            return(
                    <div className={prefixCls} ref="scroll">
                        {typeList.map((item,index)=>{
                            return(
                                <div key={index} className={this.state.activeType === index?prefixCls+"-item-active":prefixCls+"-item"} onClick={(e)=>this.onTypeClick(index,item,e)}>{item.name}</div>
                            )
                        })}
                    </div>
            )
        }
```

```
    }

    /*defaultProps.js*/
    function noop() {}
    export const defaultProps = {
        prefixCls:'s-type',
        typeList:[],
        onTypeClick:noop,
    };

    /*index.less*/
    @import '../default';
    @prefixCls:s-type;

    .@{prefixCls} {
      overflow:scroll;
      padding:@size-20 0;
      display:flex;
      align-items:center;
      .@{prefixCls}-item-active {
        color:@bc-main;
        font-weight:bold;
        min-width:50px;
        font-size:@size-16;
      }
      .@{prefixCls}-item {
        color:@bc-black;
        min-width:50px;
      }
    }
```

8.3.10 Address 组件

Address 组件主要用来封装地址选择器。

组件说明：

实现地址级联选择。

效果展示：

（1）选择省份，如图 8-12 所示。

（2）选择城市，如图 8-13 所示。

（3）选择县区，如图 8-14 所示。

（4）选择街道，如图 8-15 所示。

图 8-12

图 8-13

图 8-14

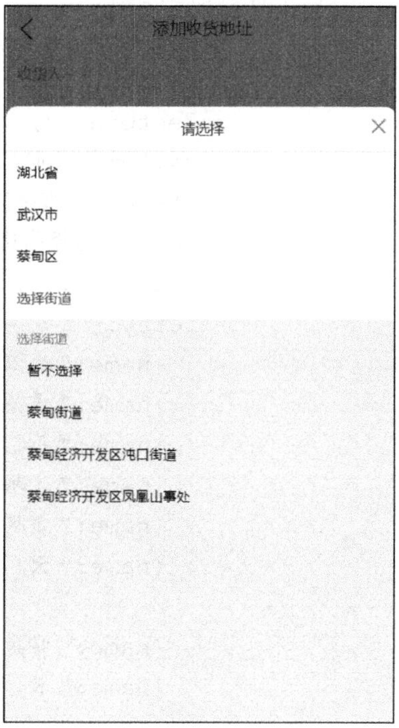

图 8-15

API 属性说明如表 8-11 所示。

表 8-11 属性说明

属性	说明	类型	默认值
closeModel	关闭地址选择框	Function	无
getAddress	获取地址信息	Function	无

index.js 内容如下：

```
import React from 'react';
import {defaultProps} from './defaultProps';
import './index.css';
import close from '../images/address/close-gray.png';
import checked from '../images/address/checked.png';
import {addressInfo} from './addressInfo';
export default class Address extends React.Component {
    static defaultProps = defaultProps;
    constructor(props){
        super(props);
        this.selectList = [];
        this.addressList = [];
        this.generate(addressInfo,this.addressList,"0");
        this.state={
            defaultText:"选择省份/城市",
            province:"",
            selectOne:"",
            city:"选择城市",
            district:"选择区县",
            street:"选择街道",
            selectList:[],
            hotCity:[
                {name:"武汉市",id:"11",superId:"1",type:"city"},
                {name:"孝感市",id:"12",superId:"1",type:"city"},
                {name:"荆门市",id:"13",superId:"1",type:"city"},
                {name:"十堰市",id:"14",superId:"1",type:"city"},
                {name:"黄冈市",id:"15",superId:"1",type:"city"},
                {name:"天门市",id:"16",superId:"1",type:"city"},

                {name:"常德市",id:"17",superId:"2",type:"city"},
                {name:"长沙市",id:"18",superId:"2",type:"city"},
                {name:"郴州市",id:"19",superId:"2",type:"city"},
```

```
                {name:"衡阳市",id:"20",superId:"2",type:"city"},
                {name:"怀化市",id:"21",superId:"2",type:"city"},
                {name:"娄底市",id:"22",superId:"2",type:"city"},
            ],
            isChoose:false,
            addressList:this.addressList
        }
    }
    // 处理数据结构，将数组处理成链表形式
    generate = (list,arr,superId)=>{
        var data = arr||[];
        for(var i =0;i<list.length;i++){
            var item = list[i];
            if(item.superId == superId){
                if(item.type == "province"){
                    item.text = " 选择城市 ";
                }
                if(item.type == "city"){
                    item.text = " 选择县区 ";
                }
                if(item.type == "district"){
                    item.text = " 选择街道 ";
                }
                data.push(item);
                data.map((item1)=>{
                    item1.list = [];
                    return this.generate(list,item1.list,item1.id);
                })
            }
        }
        return data;
    };
    // 选择热门城市
    chooseCity(text){
        var province = {};
        var item = {};
        for(var i=0;i<this.addressList.length;i++){
            var itemOne = this.addressList[i];
            // 同等类型（都是省份或者都是城市）添加后替换原先的地区，清除后面的选项
            for(var j=0;j<itemOne.list.length;j++){
```

```
                var itemOne1 = itemOne.list[j];
                if(itemOne1.type == text.type && itemOne1.name.indexOf
(text.name)>-1){
                    province = itemOne;
                    item = itemOne1;
                    break;
                }
            }
        }
        this.selectList.push(province);
        this.selectList.push(item);
        this.selectList.push({name:item.text,list:item.list});
        this.setState({
            selectOne:{name:item.text,list:item.list},
            isChoose:true,
            addressList:item.list,
            defaultText:item.text,
            selectList:this.selectList,
        })
    }
    // 改变地址
    changeAddress(item){
        this.setState({
            selectOne:item
        });
        // 找到对应的数据集
        this.findOne(this.addressList,item)
    }
    findOne(list,item){
        if(list && list.length){
            for(var i=0;i<list.length;i++){
                var item1 = list[i];
                if(item.name.indexOf("选择")>-1){
                    this.setState({
                        addressList:item.list
                    });
                    break;
                }else if(item.name == item1.name){
                    this.setState({
                        addressList:list
                    });
```

```
                break;
            }else {
                this.findOne(item1.list,item)
            }
        }
    }
}
// 选择地区
chooseAddress(item){
    let {getAddress} = this.props;
    for(var i=0;i<this.selectList.length;i++){
            var itemOne = this.selectList[i];
            // 将选择列表中带有选择的去掉
            if(itemOne.name.indexOf(" 选择 ")>-1){
                this.selectList.splice(i,1);
                break;
            }
            // 同等类型（都是省份或者都是城市）添加后替换原先的地区，清除后面的选项。
            if(itemOne.type == item.type){
                this.selectList[i] = item;
                this.selectList = this.selectList.slice(0,i+1);
                this.selectList.push({name:item.text,list:item.list});
                break;
            }
    }
    // 去重，只添加没有添加的地址
    if(!this.selectList.includes(item)){
        this.selectList.push(item);
        this.selectList.push({name:item.text,list:item.list});
    }
    this.setState({
        selectOne:{name:item.text,list:item.list},
        isChoose:true,
        addressList:item.list,
        defaultText:item.text,
        selectList:this.selectList,
    },()=>{
        if(item.list.length == 0){
            // 过滤带有选择的值
```

```
                    this.state.selectList = this.state.selectList.
filter((item)=>item.name && item.name.indexOf("选择")<0);
                    var obj = {};
                    obj.province = this.state.selectList[0]?this.state.
selectList[0].name:"";
                    obj.city = this.state.selectList[1]?this.state.
selectList[1].name:"";
                    obj.district = this.state.selectList[2]?this.state.
selectList[2].name:"";
                    obj.street = this.state.selectList[3]?this.state.
selectList[3].name:"";
                    getAddress(obj);
                }
            })
        }
        render(){
            let {closeModel,prefixCls} = this.props;
            return(
                <div className={prefixCls}>
                    <div className={prefixCls+"-wrapper"}>
                        <div className={prefixCls+"-title"}>
                            <div className={prefixCls+"-title-left"}></div>
                            <div className={prefixCls+"-title-text"}>请选择
</div>
                            <img src={close} alt="" className={prefixCls+"-
title-right"} onClick={()=>closeModel()}/>
                        </div>
                        {this.state.isChoose?
                            <div className={prefixCls+"-already"}>
                                {this.state.selectList.map((item,
index)=>{
                                    return <div key={index}
                                        className={item.
name == this.state.selectOne.name?prefixCls+"-already-item-active":
prefixCls+"-already-item"}
                                        onClick={()=>this.
changeAddress(item)}
                                    >{item.name}</div>
                                })}
                            </div>:
                            <div className={prefixCls+"-city"}>
```

```jsx
                            <div className={prefixCls+"-city-text"}>热门城市 </div>
                            <div className={prefixCls+"-city-list"}>
                                {
                                    this.state.hotCity.map((item,index)=>{
                                        return <div key={index} className={prefixCls+"-city-item"} onClick={()=>this.chooseCity(item)}>{item.name}</div>
                                    })
                                }
                            </div>
                        }
                        <div className={prefixCls+"-select"}>
                            <div className={prefixCls+"-select-text"}>{this.state.defaultText}</div>
                            <div className={prefixCls+"-select-list"}>
                                {
                                    this.state.addressList.map((item,index)=>{
                                        return <div className={prefixCls+"-select-wrapper"} key={index}>
                                            <div onClick={()=>this.chooseAddress(item,index)}
                                                className={item.name == this.state.selectOne.name?prefixCls+"-select-item-active":prefixCls+"-select-item"}>{item.name}</div>
                                            {item.name == this.state.selectOne.name? <img src={checked} alt="" className={prefixCls+"-select-item-checked"}/>:null}
                                        </div>
                                    })
                                }
                            </div>
                        </div>
                    </div>
                )
            }
        }
```

defaultProps.js 内容如下：

```js
function noop() {}
export const defaultProps = {
    prefixCls:'s-address',
    closeModel:noop,
    getAddress:noop,
};
```

addressInfo.js 内容如下：

```js
export const addressInfo =  [
    {name:" 湖北省 ",id:"1",superId:"0",type:"province"},
    {name:" 湖南省 ",id:"2",superId:"0",type:"province"},

    {name:" 武汉市 ",id:"11",superId:"1",type:"city"},
    {name:" 孝感市 ",id:"12",superId:"1",type:"city"},
    {name:" 荆门市 ",id:"13",superId:"1",type:"city"},
    {name:" 十堰市 ",id:"14",superId:"1",type:"city"},
    {name:" 黄冈市 ",id:"15",superId:"1",type:"city"},
    {name:" 天门市 ",id:"16",superId:"1",type:"city"},

    {name:" 常德市 ",id:"17",superId:"2",type:"city"},
    {name:" 长沙市 ",id:"18",superId:"2",type:"city"},
    {name:" 郴州市 ",id:"19",superId:"2",type:"city"},
    {name:" 衡阳市 ",id:"20",superId:"2",type:"city"},
    {name:" 怀化市 ",id:"21",superId:"2",type:"city"},
    {name:" 娄底市 ",id:"22",superId:"2",type:"city"},

    {name:" 蔡甸区 ",id:"23",superId:"11",type:"district"},
    {name:" 东西湖区 ",id:"24",superId:"11",type:"district"},
    {name:" 汉南区 ",id:"25",superId:"11",type:"district"},

    {name:" 暂不选择 ",id:"26",superId:"23",type:"street"},
    {name:" 蔡甸街道 ",id:"27",superId:"23",type:"street"},
    {name:" 蔡甸经济开发区沌口街道 ",id:"28",superId:"23",type:"street"},
    {name:" 蔡甸经济开发区凤凰山事处 ",id:"29",superId:"23",type:"street"},
];
```

index.less 内容如下:

```less
@prefixCls:s-address;
.@{prefixCls} {
  position:fixed;
  left:0;
  right:0;
  bottom:0;
  top:0;
  z-index:999;
  background:rgba(0,0,0,0.5);
  .@{prefixCls}-wrapper{
    margin-top:25%;
    background:@bc-white;
    border-radius:@size-10;
    .@{prefixCls}-title{
      display:flex;
      border-bottom:1px solid @c-eee;
      justify-content:space-between;
      padding:@size-10;
      .@{prefixCls}-title-left,.@{prefixCls}-title-right{
        width:(@size-20)-5;
        height:(@size-20)-5;
      }
      .@{prefixCls}-title-text{
        color:@c-333;
      }
    }
    .@{prefixCls}-city {
      padding-top:@size-10;
      margin:0 @size-10;
      .@{prefixCls}-city-text {
        color:@c-999;
        font-size:@size-12;
      }
      .@{prefixCls}-city-list {
        .@{prefixCls}-city-item {
          display:inline-block;
          width:25%;
          text-align:center;
```

```less
      padding:@size-10 0;
      font-size:@size-12+1;
    }
  }
}
.@{prefixCls}-already {
  margin:0 @size-10;
    .@{prefixCls}-already-item {
      padding:@size-10 0;
      font-size:@size-12+1;
    }
   .@{prefixCls}-already-item-active {
      padding:@size-10 0;
      font-size:@size-12+1;
      color:@c-main;
    }
}
.@{prefixCls}-select {
  padding:@size-10;
  min-height:100vh;
  background:@bc-body;
  .@{prefixCls}-select-text {
    color:@c-999;
    font-size:@size-12;
  }
  .@{prefixCls}-select-list {
    .@{prefixCls}-select-wrapper{
      display:flex;
      justify-content:space-between;
      align-items:center;
      .@{prefixCls}-select-item {
        padding:@size-10;
        font-size:@size-12+1;
      }
      .@{prefixCls}-select-item-active {
        padding:@size-10;
        font-size:@size-12+1;
        color:@c-main;
      }
      .@{prefixCls}-select-item-checked {
        width:(@size-20)-5;
```

```
                height:(@size-20)-5;
            }
        }
    }
   }
 }
}
```

8.4 界面分析

在实际开发之前，为了提高效率，工程师应该首先对照着 UI 设计图对界面进行简单分析，这个过程可以记录下来，也可以在心里有个大概的轮廓，这样有利于后期开发的时候减少一些不必要的错误和沟通成本。而不是盲目地一边开发一边去思考这些问题，做任何事情之前有一个总体上的规划将会事半功倍。那么在页面开发之前需要思考哪些方面的问题呢？

8.4.1 页面结构

页面结构指的是整个页面布局可以分为哪几个部分，每一部分应该如何开发；是使用 antd-mobile 组件库现有的组件，还是使用自己封装的公共组件，抑或是直接编写。

8.4.2 图片元素

分析界面中的图片有哪几种，动态还是静态，纯图片还是图标，然后将界面中所需要的静态图片资源（包括图标）添加到对应的文件中。

8.4.3 页面色彩

一般应用中都会有一个主题背景色和字体颜色。而且为了使界面不那么花哨，除了图标或图片之外，一个界面色彩不会太多。本项目中的主题背景色和字体颜色为红色，颜色值为 #cc0000；边框和下划线采用淡灰色，颜色值为 #cccccc；界面主体部分字体采用黑色，颜色值为 #000000；非主体部分字体采用灰色，颜色值为 #999999。

8.4.4 页面边距

在实际开发中详细的页面边距可以跟设计师沟通，也可以自己根据视觉调整。因为浏览器的兼容问题以及显示屏色差问题，往往真正开发出来的效果与设计师画出来的效果图会有所出入，所以最好的做法是一边跟设计师沟通，一边根据实际情况灵活调整。

本项目中使用 rem 单位。rem 是指相对于根元素的字体大小的单位，即表示所有节点与根节点 HTML 字体大小的相对值。Chrome 浏览器默认的字体大小是 16px，即在没有特意设置 HTML 字体大小的情况下，1rem = 1×16px，2rem = 2×16px = 32px，所以 rem 单位可以随着 HTML 字体大小的设置而变化。比如设置 HTML 字体大小为 20px，那么 1rem 就等于 20px。这里为了换算关系更容易计算，将 HTML 字体大小设置为 62.5%，也就相当于 10px。后期使用 rem 单位时只需要记住 rem 是 px 的 10 倍，即 3rem 就是 30px。界面中的上、下边距一般随业务场景不同会有较大差异，而左、右边距一般都是固定的，本项目中的左、右边距统一为 10px。

第 9 章　用户模块开发

【本章导读】

◎ 开发注册页面

◎ 开发登录页面

9.1　注册页面开发

1. 页面入口

（1）点击一级页面【我的】，未登录状态时，进入登录页面。

（2）点击【快速注册】，进入注册页面。

2. 页面示例

效果如图 9-1 所示。

3. 页面结构

本页面可分为 5 个部分。

第 1 部分：标题部分，可使用公共组件 Title。

第 2 部分：选择图片部分，没有公共组件，自由编写。

第 3 部分：文本框部分，可使用公共组件 List+Input。

第 4 部分：【注册】按钮部分，可使用公共组件 Button。

第 5 部分：点击【已有账户？登录】进入登录页面，自由编写。

图 9-1

4. 页面数据

点击【注册】，调取注册接口 register。

5. 编写代码

（1）创建模块及文件。

在 src 目录下新建 pages 目录（后面所有的业务组件都将放在 pages 目录下），在 pages 目录下新建 register 目录，在 register 目录下新建 register.js 文件和 register.css 文件。

在 src 目录下新建 images 目录，在 images 目录下分别新建 index 目录、login 目录、register 目录以及 personal 目录，并分别将对应的图片放进去。

（2）配置路由

在 src 目录下新建 router.js 文件作为路由配置，并添加注册页面路由，代码如下。

```
import React from 'react';
import {HashRouter, BrowserRouter,Route,Redirect,Switch,Link,NavLink} from 'react-router-dom';
import Register from './pages/register/register';
const Router = () => (
    <BrowserRouter>
<Route exact  path="/register"  component={Register}></Route>
    </BrowserRouter>
);
export default Router;
```

（3）register.js 文件的代码如下。

```
import React from 'react';
import './register.css';
import {Title,List,Input,Button} from '../../share';
import {getData, goNext,isDefine,checkParam} from '../../utils';
import {Toast} from 'antd-mobile';
import verifyCode from '../../images/register/verify-code.png';
export default class Register extends React.Component {
    constructor(props) {
        super(props);
        this.state = {
            userAvatar:[],
            getCode:'获取验证码',
            userName:'',
```

```
            mobile:'',
            verifyCode:"",
            password:'',
            confirmPwd:'',
        }
    }
    change(state){
        this.setState(state)
    }
    getVerify(){
        if(!isDefine(this.state.mobile)){
            Toast.info("请输入手机号码");
            return;
        }
        getData({
            method:'post',
            url:'getVerifyCode',
            data:{
                mobile:this.state.mobile,
                type:1
            },
            successCB:(res)=> {
                this.setState({
                    getCode:60
                },()=>{
                    this.timer = setInterval(()=>{
                        // 验证码过了60秒，消除定时器
                        if(this.state.getCode == 1){
                            clearInterval(this.timer);
                            this.setState({
                                getCode:"获取验证码"
                            })
                        }else {
                            // 验证码60秒倒计时
                            this.setState({
                                getCode:--this.state.getCode
                            })
                        }
                    },1000)
                })
            }
```

```jsx
            })
        }
        renderRight(){
            return(
                <div className="cm-flex cm-ai-c"><img src={verifyCode} alt="" className="cm-img-04"/><span className="cm-c-main" onClick={()=>this.getVerify()}> {this.state.getCode==" 获取验证码 "?" 获取验证码 ":this.state.getCode+" 秒后重新获取 "}</span></div>
            )
        }
        register(){
            // 判断参数是否为空
            var arr = [
                {value:this.state.userAvatar,msg:" 请选择头像 "},
                {value:this.state.userName,msg:" 请输入昵称 "},
                {value:this.state.mobile,msg:" 请输入手机号码 "},
                {value:this.state.verifyCode,msg:" 请输入验证码 "},
                {value:this.state.password,msg:" 请输入密码 "},
                {value:this.state.confirmPwd,msg:" 请输入确认密码 "},
                {value:this.state.password == this.state.confirmPwd,msg:" 密码与确认密码不一致 "},
            ];
            checkParam(arr,()=> {
                let params = new FormData(); // 创建 form 对象，通过 append 向 form 对象添加数据
                params.append('userAvatar',this.state.userAvatar[0]);
                params.append('userName',this.state.userName);
                params.append('mobile',this.state.mobile);
                params.append('verifyCode',this.state.verifyCode);
                params.append('password',this.state.password);
                params.append('confirmPwd',this.state.confirmPwd);
                getData({
                    method:'post',
                    url:'register',
                    data:params,
                    successCB:(res) => {
                        // 注册成功之后调用登录接口
                        getData({
                            method:'post',
                            url:'login',
                            data:{
```

```
                                mobile:this.state.mobile,
                                password:this.state.password
                            },
                            successCB:(res)=> {
                                // 登录成功之后将个人信息缓存至本地
                                localStorage.setItem("userInfo",JSON.stringify(res.result));
                                localStorage.setItem("token",res.result.token);
                                // 进入首页
                                goNext(this,'index')
                            }
                        })
                    }
                })
            })
        }
        // 获取头像
        getAvatar(event){
            var files = event.target.files;
            this.setState({
                userAvatar:files
            });
            if(window.FileReader) {
                var file = files[0];
                var fr = new FileReader();
                fr.onloadend = (e)=> {
                    this.refs.avatar.style.backgroundImage  = "url("+e.target.result+")";
                    this.refs.avatar.style.backgroundRepeat = "no";
                    this.refs.avatar.style.backgroundSize = "100%";
                };
                // 将图片作为 URL 读出
                fr.readAsDataURL(file);
            }
            // 添加完头像后将伪元素 :before 和 :after 去掉, 也就是图片中的 + 去掉
            this.refs.avatar.setAttribute("class","register-again");
        }
        render() {
            return<div>
                <Title that={this}  title=" 注册 "/>
```

```jsx
                    <div className="cm-mt-036 register-relative">
                        <div className="register-input-box" ref="avatar">
                            <input type="file" className="register-input" onChange={(e)=>this.getAvatar(e)}/>
                        </div>
                    </div>
                    <div className="cm-mlr-018 cm-mt-036">
                        <List>
                            <Input
                                placeholder=" 请输入昵称 "
                                value={this.state.userName}
                                onChange={(val)=>this.change({userName:val})}
                                maxLength={16}
                            />
                        </List>
                        <List>
                            <Input
                                placeholder=" 请输入手机号码 "
                                value={this.state.mobile}
                                onChange={(val)=>this.change({mobile:val})}
                                maxLength={11}
                            />
                        </List>
                        <List>
                            <Input
                                placeholder=" 请输入验证码 "
                                value={this.state.verifyCode}
                                onChange={(val)=>this.change({verifyCode:val})}
                                renderRight = {()=>this.renderRight()}
                                maxLength={6}
                            />
                        </List>
                        <List>
                            <Input
                                type="password"
                                placeholder=" 请输入密码 "
                                value={this.state.password}
                                onChange={(val)=>this.change({password:val})}
                                maxLength={16}
                            />
                        </List>
```

```
                <List>
                    <Input
                        type="password"
                        placeholder=" 请输入确认密码 "
                        value={this.state.confirmPwd}
                        onChange={(val)=>this.change({confirmPwd:val})}
                        maxLength={16}
                    />
                </List>
                <Button type="fill" className="cm-mt-080" onClick=
{() => this.register()}>注册 </Button>
                <div className="cm-tx-r login-operate" onClick={()=
>goNext(this,"login")}>
                        已有账户？登录
                </div>
            </div>
        </div>
    }
}
```

（4）register.css 文件的代码如下。

```
.register-relative{
    position:relative;
}
.register-input{
    width:1.6rem;
    height:1.6rem;
    opacity:0;
}
.register-again,.register-input-box{
    width:1.6rem;
    border-radius:50%;
    background:#ffffff;
    border:1px solid #ddd;
    margin:auto;
    height:1.6rem;
    color:#ddd;
}
.register-input-box:before,.register-input-box:after{
```

```
        content:"";
        height:0.5rem;
        width:0.05rem;
        background:#ddd;;
        position:absolute;
        top:50%;
        left:50%;
        transform:translate(-50%, -50%)
}
.register-input-box:after{
    height:0.05rem;
    width:0.5rem;
}
```

9.2 登录页面开发

1. 页面入口

点击一级页面【我的】,未登录状态时,进入登录页面。

2. 页面示例

效果如图 9-2 所示。

3. 页面结构

本页面可分为 5 个部分。

第 1 部分：标题部分,可使用公共组件 Title。

第 2 部分：图片部分,自由编写。

第 3 部分：文本框部分,可使用公共组件 List+Input。

第 4 部分：【登录】按钮部分,可使用公共组件 Button。

第 5 部分：跳转部分,自由编写。

4. 页面数据

点击【登录】,调取登录接口 login。

5. 编写代码

(1) 创建模块及文件。

在 pages 目录下新建 login 目录, 在 login 目录下

图 9-2

新建 login.js 文件和 login.css 文件。

（2）配置路由。

在 router.js 文件中配置商品详情页路由，代码如下。

```
import Login from './pages/login/login';
<Route exact  path="/login"  component={Login}></Route>
```

（3）login.js 文件的代码如下。

```
import React from 'react';
import './login.css';
import account from '../../images/login/account.png';
import password from '../../images/login/password.png';
import defaultAvatar from '../../images/login/avatar.jpg';
import {Title, List, Input,Button} from '../../share';
import {getData, goNext, checkParam} from '../../utils';
import {Toast} from 'antd-mobile';
export default class Publish extends React.Component {
    constructor(props) {
        super(props);
        this.state = {
            account:'',
            password:'',
        }
    }
    change(state) {
        this.setState(state)
    }
    // 进入注册页面
    goRegister() {
        goNext(this, 'register');
    }
    login() {
        var arr = [
            {value:this.state.mobile, msg:"请输入手机号码"},
            {value:this.state.password, msg:"请输入密码"},
        ]
        checkParam(arr, () => {
            getData({
```

```jsx
                    method:'post',
                    url:'login',
                    data:{
                        mobile:this.state.mobile,
                        password:this.state.password,
                    },
                    successCB:(res) => {
                        var token = res.result.token;
                        // 登录成功之后将个人信息缓存到本地，并进入首页
                        localStorage.setItem("userInfo", JSON.stringify(res.result));
                        localStorage.setItem("token", token);
                        goNext(this, 'index')
                    }
                })
            })
        }

        render() {
            return (
                <div>
                    <Title that={this} title=" 登录 "/>
                    <div className="cm-tx-c cm-mt-036"><img src={defaultAvatar} alt="" className="cm-img-16 cm-border-radius-half"/></div>
                    <div className="cm-mlr-018 cm-mt-036">
                        <List>
                            <Input
                                src={account}
                                placeholder=" 请输入账号 "
                                value={this.state.mobile}
                                onChange={(val) => this.change({mobile:val})}
                                style={{textAlign:'left'}}
                            />
                        </List>
                        <List>
                            <Input
                                type="password"
                                src={password}
                                placeholder=" 请输入密码 "
                                value={this.state.password}
```

```
                            onChange={(val) => this.change
({password:val})}
                            style={{textAlign:'left'}}
                    />
                </List>
                <Button type="fill" className="cm-mt-080" onClick=
{() => this.login()}>登录</Button>
                <div className="cm-tx-c login-operate">
                    <span onClick={() => this.goRegister()}>快速
注册</span>
                    <span>|</span>
                    <span onClick={()=>goNext(this,"forgotPwd")}>
忘记密码?</span>
                </div>
            </div>
        </div>)
    }
}
```

(4) login.css 代码如下。

```
.login-operate{
    margin:0.36rem auto;
    display:flex;
    justify-content:space-around;
    width:40%;
    color:#cc0000;
}
```

第 10 章 商品模块开发

【本章导读】

◎ 开发发布 / 修改商品页面

◎ 开发商品列表 / 首页

◎ 开发商品详情页面

10.1 发布 / 修改商品页面开发

图 10-1

1. 页面入口

点击一级页面【发布】，进入商品发布页面。

2. 页面示例

效果如图 10-1 所示。

3. 页面结构

本页面可分为 6 个部分。

第 1 部分：标题部分，可使用公共组件 Title。

第 2 部分：商品描述部分，自由编写。

第 3 部分：选择图片部分，可使用 antd-mobile 中的 ImagePicker 组件。

第 4 部分：文本框部分，可使用公共组件 List + Input + Select。

第 5 部分：【发布】按钮部分，可使用公共组件 Button。

第 6 部分：底部 tab 切换部分，可使用公共组件 TabBottom。

4. 页面数据

点击【发布】，调取商品发布接口 publishProduct。

5. 编写代码

（1）创建模块及文件。

在 pages 目录下新建 publish 目录，在 publish 目录下新建 publish.js 文件和 publish.css 文件。

（2）配置路由。

在 router.js 文件中配置商品详情页路由，代码如下。

```
import Publish from './pages/publish/publish';
<Route exact path="/publish" component={Publish}></Route>
```

（3）publish.js 文件的代码如下。

```
import React from 'react';
import {ImagePicker,Toast} from 'antd-mobile';
import './publish.css';
import {Select,Title,List,Input,Button,TabBottom} from '../../share';
import {getData,goNext,checkParam,isDefine} from '../../utils';
export default class Publish extends React.Component {
    constructor(props) {
        super(props);
        this.state = {
            productImgs:[],
            typeList:[],
            productType:'',
            productPrice:0,
            productDesc:'',
        };
    }
    componentDidMount(){
        // 检查登录状态，如果未登录，在 getData 方法中统一处理未登录，进入登录页面
        getData({
            method:'post',
            url:'checkLoginValid',
        });
        this.getTypeList();
    }
    // 获取商品类别 id
```

```
changeType(item){
    this.setState({
        productType :item.id
    })
}
onChange = (productImgs, type, index) => {
    this.setState({
        productImgs,
    });
}
// 获取商品描述
changeDes(event){
    this.setState({
        productDesc:event.target.value
    })
}
// 金额正则校验
changePrice(event){
    event.target.value = event.target.value.replace(/[^\d.]/g,"");
    // 清除数字和"."以外的任意字符
    event.target.value = event.target.value.replace(/\.{2,}/g,".");
    // 只保留第一个".", 清除多余连续的"."
    event.target.value = event.target.value.replace(".","$#$").replace(/\./g,"").replace("$#$",".");
    // 将第一个"."用特殊符号替换, 其他"."都转换为空, 再将特殊符号用"."
    替换回来,目的是保证清除非连续的"."
    event.target.value = event.target.value.replace(/^(\-)*(\d+)\.(\d\d).*$/,'$1$2.$3');
    // 只能输入两个小数
    if(event.target.value.indexOf(".")< 0 && event.target.value !=""){
    // 以上已经过滤,此处控制的是,如果没有小数点,首位不能为类似于 01、02 的金额
        event.target.value= parseFloat(event.target.value);
    }
    // 通过 setState 改变金额
    this.setState({
        productPrice:event.target.value
    })
}
// 首页获取了商品类型列表并将其放入缓存,这里直接读取缓存
getTypeList(){
    if(isDefine(localStorage.getItem("typeList"))){
```

```js
            this.setState({
                typeList:JSON.parse(localStorage.getItem("typeList"))
            })
        }
    }
    submit(){
        var arr = [
            {value:(this.state.productDesc.length>9||this.state.productDesc.length<300), msg:"商品描述请输入10～300个字符"},
            {value:this.state.productDesc, msg:"请输入商品描述"},
            {value:this.state.productImgs[0], msg:"请添加图片"},
            {value:this.state.productPrice, msg:"请输入商品价格"},
            {value:this.state.productType, msg:"请选择分类"},
        ];
        checkParam(arr,()=>{
            var token = localStorage.getItem("token");
            let param = new FormData(); // 创建form对象，通过append向form对象添加数据
            for (var i = 0; i < this.state.productImgs.length; i++){
                param.append('productImgs',this.state.productImgs[i].file);// 将每一张图片添加进去
            }
            param.append('productDesc',this.state.productDesc);
            param.append('productPrice',this.state.productPrice);
            param.append('productTypeId',this.state.productType);
            getData({
                method:'post',
                url:'publishProduct',
                headers:{
                    'Content-Type':'multipart/form-data',
                    'token':token
                },
                data:param,
                successCB:(res) => {
                    console.log("发布接口成功"+new Date());
                    goNext(this,"index");
                }
            })
        })
    }
    change(state){
```

```jsx
            this.setState(state)
    }
    render(){
        const { productImgs } = this.state;
        return(
            <div>
                <Title isHome={true} title=" 商品发布 "/>
                <div className="cm-mlr-018 cm-mt-018">
                    <textarea name="" id="" cols="30"
                            onChange={(e)=>this.changeDes(e)}
                            maxLength="300"
                            defaultValue={this.state.productDesc}
                            className="cm-w-full cm-fs-026 cm-
border-ddd cm-p-018 publish-textarea" rows="10" placeholder=" 描述商品转手
原因、入手渠道和使用感受 "></textarea>
                    <ImagePicker
                        files={productImgs}
                        onChange={this.onChange}
                        onImageClick={(index, fs) => console.log
(index, fs)}
                        selectable={productImgs.length < 9}
                    />
                </div>
                <div className="cm-mlr-018">
                <List>
                    <Input
                        label=" 价格 "
                        style = {{textAlign:"right"}}
                        placeholder=" 请输入商品价格 "
                        onChange={(val)=>this.change({productPrice:val})}
                        maxLength={11}
                    />
                </List>
                <List>
                    <Select
                        options={this.state.typeList}
                        babel=" 分类 "
                        onChange={this.changeType.bind(this)}
                    />
                </List>
                </div>
```

```
                    <Button className="cm-mt-080" onClick={()=>this.
submit()}>发布</Button>
                    <TabBottom history={this.props.history}
activeNum={1}/>
                </div>
            )
        }
    }
```

(4) publish.css 文件的代码如下。

```
.publish-textarea::-webkit-input-placeholder{
    height:120px;
    line-height:120px;
    text-align:center;
    box-sizing:border-box;
}
.am-image-picker-list{
    padding:9px 9px 8px 0!important;
}
```

10.2 商品列表 / 首页开发

1. 页面入口

点击一级页面【首页】，进入首页。

2. 页面示例

效果如图 10-2 所示。

3. 页面结构

本页面可分为 5 个部分。

第 1 部分：文本框与扫码部分，可使用公共组件 Search。

第 2 部分：轮播图部分，使用 antd-mobile 中的 Carousel 组件。

第 3 部分：分类部分，可使用公共组件 Type。

图 10-2

第 4 部分：商品部分，使用 antd-mobile 中的 ListView 组件。

第 5 部分：底部 tab 切换部分，可使用公共组件 TabBottom。

4. 页面数据

（1）页面初始化时调取轮播图接口 getBannerList。

（2）页面初始化时调取商品列表接口 getProductList。

（3）页面初始化时调取商品分类接口 getProductTypeList。

5. 编写代码

（1）创建模块及文件。

在 src/pages 目录下，新建 index 目录，在 index 目录下新建 index.js 文件用来编写首页内容，处理业务逻辑；新建 index.css 文件用来编写首页所需要的样式。

（2）配置路由。

在 router.js 文件中配置首页路由，因为首页是默认页，所以需要配置 / 指向首页，代码如下。

```
import Index from './pages/index/index';
<Route exact  path="/"  component={Index}/>
<Route exact  path="/index"  component={Index}/>
```

（3）index.js 文件的代码如下。

```
import React from 'react';
import publish from '../../images/index/publish.png';
import {Carousel,ListView} from 'antd-mobile';
import {Search,Type,TabBottom} from '../../share';
import {getData,goNext} from '../../utils';
import './index.css';
export default class Index extends React.Component {
    constructor(props){
        super(props);
        this.initData = [];// 定义空数组，用于上拉加载更多时拼接数据
        this.typeId="";// 定义商品类型查询
        this.key="";// 定义关键字查询
        const dataSource = new ListView.DataSource({
            rowHasChanged:(row1, row2) => row1 !== row2,
        })// 数据源初始化
```

```
            this.state = {
                //isLoading:true,// 用来控制上拉加载更多时是否显示 loading
                dataSource:dataSource.cloneWithRows(this.initData),
                pageNum:1,// 定义初始化请求序号
                pageSize:10,// 定义初始化请求数量
                typeList:[{name:" 全部 "}],// 定义初始化商品类型数组
                productList:[],// 定义初始化商品数组
                hasMore:true,// 是否已经加载完成
                bannerList:[],// 定义初始化轮播图数组
                statusText:" 加载中 "// 定义上拉加载判断状态，3 种状态：加载中、加载完成、暂无数据
            }
        }
        // 选择商品类型
        changeType(item){
            this.refs.search.closeValue();
            this.typeId = item.id;
            this.state.pageNum = 1;
            this.key = "";
            this.initData = [];
            this.setState({
                dataSource:this.state.dataSource.cloneWithRows(this.initData),
                //isLoading:false,
                hasMore:true,
                statusText:" 加载中 "
            });
            this.getProductList();
        }
        // 点击某一商品图片进入商品详情页面
        goDetail(item){
            goNext(this,"indexDetail",item);
        }
        // 点击进入商品发布页面
        goPublish(e){
            goNext(this,"publish");
        }
        // 获取商品列表
        getProductList(){
            getData({
```

```
                    method:'post',
                    url:'getProductList',
                    isShowLoad:true,
                    data:{
                        pageNum:this.state.pageNum,
                        pageSize:this.state.pageSize,
                        productDesc:this.key,
                        productTypeId:this.typeId
                    },
                    successCB:(res) => {
                        this.state.pageNum++;
                        this.initData = this.initData.concat(res.result.list);// 已有的数组拼接上拉加载更多的数组
                        let total = res.result.total;// 商品列表总数量
                        if(total == 0){
                            this.setState({
                                statusText:"暂无数据"// 总数为 0 时表示没有数据
                            })
                        }else if(total == this.initData.length){
                            this.setState({
                                dataSource:this.state.dataSource.cloneWithRows(this.initData),
                                hasMore:false,// 加载完成，hasMore 状态改为 false
                                statusText:" 加载完成 "// 拼接的数组长度等于总数时表示加载完成
                            })
                        }else {
                            setTimeout(()=>{
                                this.setState({
                                    dataSource:this.state.dataSource.cloneWithRows(this.initData),
                                    //isLoading:false,
                                });
                            },600)

                        }
                    }
                })
            }
            // 获取轮播图列表
```

```
        getBannerList(){
            getData({
                    method:'get',
                    url:'getBannerList',
                    successCB:(res) => {
                        this.setState({
                            bannerList:res.result
                        })
                    }
            })
        }
        // 获取商品类型列表
        getProductTypeList(){
            getData({
                method:'get',
                url:'getProductTypeList',
                successCB:(res) => {
                    localStorage.setItem("typeList",JSON.stringify(res.result));
                    let typeList = this.state.typeList.concat(res.result);
                    this.setState({
                        typeList:typeList
                    })
                }
            })
        }
        // 通过商品关键字搜索
        getProductListByKey(val){
            this.state.pageNum = 1;
            this.initData = [];
            this.setState({
                dataSource:this.state.dataSource.cloneWithRows(this.initData),
                hasMore:true,
                statusText:"加载中"
            })
            this.key = val;
            this.getProductList()
        }
        // 加载页面时获取轮播图列表、商品类型列表、商品列表
        componentDidMount(){
```

```
            this.getProductList();
            this.getProductTypeList();
            this.getBannerList();
        }
        // 上拉加载更多时触发
        onEndReached = (event) => {
            if (!this.state.hasMore) {
                return false;
            }
            this.setState({ isLoading:true},()=>{
                this.getProductList();
            });
        }

        // 渲染页面
        render() {
            var arr=[];
            const row = (rowData, sectionID, rowID) => {
                var productImgs = rowData.productImgs.split(",");
                var desc = rowData.productDesc && rowData.productDesc.slice(0,28);
                return (
                        <div className={rowID%2==0?"index-product-left":"index-product-right"} key={rowID} onClick={()=>this.goDetail(rowData)}>
                            <div className="index-img">
                                <img src={productImgs[0]} alt="" className="index-img"/>
                            </div>
                            <div className="cm-p-009">
                            <div className="cm-c-666 cm-mt-009 cm-fs-024">{desc}</div>
                            <div className="index-price cm-c-main cm-fs-026 cm-flex cm-jc-sb">
                                <span>￥{rowData && rowData.productPrice}</span>
                                <span className="cm-c-999 cm-fs-022">{rowData.wantNum>0?rowData.wantNum+" 人想要 ":""}</span>
                            </div>
                            <div className="cm-flex cm-ptb-009 cm-border-top-eee cm-ai-c">
                                <img src={rowData.
```

```jsx
publishUserAvatar} className="cm-img-04 cm-border-radius-half cm-mr-
009" alt=""/>
                            <div className="cm-c-333 cm-fs-026">
{rowData.publishUserName}</div>
                        </div>
                    </div>
                </div>
        );
    };
    return (
        <div>
            <Search
                leftClick = {(val)=>this.getProductListByKey(val)}
                rightIcon={publish}
                rightClick={()=>this.goPublish()}
                placeholder=" 请输入商品关键字 "
                ref="search"
            />
            <div className="cm-img-banner">
                {this.state.bannerList.length>0?
                    <Carousel>
                        {this.state.bannerList.map((item,index) => {
                            var productImgs = item.productImgs && item.productImgs.split(",");
                            return (<img src={productImgs[0]} key={index} alt="" className="cm-img-banner" onClick={()=>this.goDetail(item.id)}/>)
                        })}
                    </Carousel>:null
                }
            </div>
            <div className="cm-mlr-018">
                <Type
                    typeList = {this.state.typeList}
                    onTypeClick = {(val)=>this.changeType(val)}
                />
            </div>
            <ListView
                useBodyScroll={true}
```

```
                                    ref={el => this.lv = el}
                                    dataSource={this.state.dataSource}
                                    renderFooter={() => (<div style={{ paddingB
ottom:60, textAlign:'center' }}>
                                        {this.state.statusText}
                                    </div>)}
                                    renderBodyComponent={() => <Body/>}
                                    renderRow={row}
                                    pageSize={4}
                                    scrollRenderAheadDistance={500}
                                    onEndReached={this.onEndReached}
                                    onEndReachedThreshold={10}
                                />
                                <TabBottom history={this.props.history}/>
                </div>
            )
        }
    }
    function Body(props) {
        return(
            <div>
                <div>
                    {props.children}
                </div>
            </div>
        )
    }
```

（4）index.css 文件的代码如下。

```
.index-product-left{
    margin:0 0.09rem 0.15rem 0.18rem;
    border-radius:0.08rem;
    box-shadow:0.03rem 0.03rem 0.03rem #ddd;
    width:calc(50% - 0.27rem);
    display:inline-block;
    vertical-align:top;
}
.index-product-right{
```

```css
    margin:0 0.18rem 0.15rem 0.09rem;
    border-radius:0.08rem;
    box-shadow:0.03rem 0.03rem 0.03rem #ddd;
    width:calc(50% - 0.27rem);
    display:inline-block;
    vertical-align:top;
}
.index-img{
    border-top-left-radius:0.1rem;
    border-top-right-radius:0.1rem;
    height:3.8rem;
    width:100%;
}
.detail-line{
    display:inline-block;
    padding:0.1rem 0;
    border-bottom:1px solid #cc0000;
}
.detail-fill{
    height:0.8rem;
}
.detail-bottom{
    position:fixed;
    left:0;
    right:0;
    bottom:0;
    height:0.8rem;
    padding:0 0.18rem;
    background:#fff;
}
.detail-replay{
    margin-left:0.58rem;
}
.detail-textarea{
    border:1px solid #ddd;
    border-radius:0.1rem;
    flex:1;
    padding:0.1rem;
    margin-right:0.18rem;
}
```

10.3 商品详情页面开发

图 10-3

1. 页面入口

（1）点击一级页面【首页】，进入首页。

（2）点击商品图片，进入商品详情页面。

2. 页面示例

效果如图 10-3 所示。

3. 页面结构

本页面可分为 7 个部分。

第 1 部分：标题部分，可使用公共组件 Title。

第 2 部分：头像与昵称部分，自由编写。

第 3 部分：商品描述部分，自由编写。

第 4 部分：商品图片部分，自由编写。

第 5 部分：点赞列表部分，自由编写。

第 6 部分：留言列表部分，自由编写。

第 7 部分：点赞、留言、聊一聊部分，自由编写，需要用到绝对定位。

4. 页面数据

（1）页面初始化时调取商品详情接口 getProductDetail。

（2）页面初始化时调取评论列表接口 getCommentReplyList。

（3）页面初始化时调取点赞列表接口 getPraiseList。

（4）点击【点赞】，调取点赞接口 praiseOrUnPraise。

（5）点击【留言】，弹出留言文本框，输入留言内容，点击【发送】，调取留言接口 commentOrReply。

5. 编写代码

（1）创建模块及文件。

在 src/pages/index 目录下新建 indexDetail.js 文件用来编写商品详情页内容，处理业务逻辑。

（2）配置路由。

在 router.js 文件中配置商品详情页路由，代码如下。

```
import IndexDetail from './pages/index/indexDetail';
<Route exact path="/indexDetail" component={IndexDetail}/>
```

（3）indexDetail.js 文件代码如下。

```
import React from 'react';
import praise from '../../images/index/praise.png';
import alreadyPraise from '../../images/index/already-praise.png';
import praiseHeart from '../../images/index/praise-heart.png';
import msg from '../../images/index/msg.png';
import comment from '../../images/index/comment.png';
import {Title} from '../../share';
import {getData,getTime,autoTextarea} from '../../utils';
import './index.css';
export default class IndexDetail extends React.Component {
    constructor(props) {
        super(props);
        this.state = {
            praiseList:[],// 点赞列表
            commentList:[],// 评论列表
            isPraise:2, // 是否点赞, 1 为是, 2 为否, 默认为 2
            isShowComment:false,// 是否弹出留言框
            item:{}
        };
        this.commentItem = {};
        // 获取从首页和轮播图点击产生的参数
        var item = this.props.history.location.state ||{};
        this.id = item.id;
    }
    componentDidMount(){
        this.getProductById();
        this.getPraiseList();
        this.getCommentList();
    }

    getProductById(){
        getData({
            method:'get',
            url:'getProductDetail',
            data:{productId:this.id},
```

```js
            successCB:(res) => {
                this.setState({
                    item:res.result
                })
            }
        })
    }
    // 获取点赞列表
    getPraiseList(){
        getData({
            method:'get',
            url:'getPraiseList',
            data:{productId:this.id},
            successCB:(res) => {
                this.setState({
                    praiseList:res.result.list,
                    isPraise:res.result.praiseStatus
                })
            }
        })
    }
    // 获取评论列表
    getCommentList(){
        getData({
            method:'get',
            url:'getCommentReplyList',
            data:{productId:this.id},
            successCB:(res) => {
                this.setState({
                    commentList:res.result
                })
            }
        })
    }
    // 打开评论窗口
    showBox(e,commentItem){
        e.stopPropagation();
        // 先判断登录状态
        getData({
            method:'post',
            url:'checkLoginValid',
```

```js
            successCB:(res) => {
                this.commentItem = commentItem || this.state.item;
                this.setState({
                    isShowComment:true
                },function () {
                    //textarea 文本框高度自适应
                    autoTextarea();
                })
            }
        });
    }
    // 关闭评论窗口
    closeBox(){
        this.setState({
            isShowComment:false
        })
    }
    // 评论
    comment(){
        var userId = this.commentItem.fromUserId;
        if(this.commentItem.fromUserId == this.state.item.publishUserId){
            userId = this.commentItem.toUserId;
        }
        this.setState({
            isShowComment:!this.state.isShowComment
        });
        var value = this.refs.textarea.value;
        getData({
            method:'post',
            url:'checkLoginValid',
            successCB:(res) => {
                getData({
                    method:'post',
                    url:'commentOrReply',
                    data:{
                        productId:this.id,
                        content:value,
                        toUserId:userId,
                        replyId:this.commentItem.id,
                    },
                    successCB:(res) => {
```

```
                            this.getCommentList()
                        }
                    })
                }
            });
        }
        // 点赞
        praise(){
            getData({
                method:'post',
                url:'checkLoginValid',
                successCB:(res) => {
                        var isPraise = (this.state.isPraise == 2)?1:2;
                        this.setState({
                            isPraise:isPraise
                        });
                        getData({
                            method:'post',
                            url:'praiseOrUnPraise',
                            data:{
                                productId:this.id,
                                status:isPraise
                            },
                            successCB:(res) => {
                                this.getPraiseList();
                            }
                        })
                }
            });
        }
        // 进入聊天详情页面，初始化聊天
        goChat(){
            var param = {
                productId:this.state.item.productId,
                toUserId:this.state.item.publishUserId,
            }
            getData({
                method:'post',
                url:'initChat',
                data:param,
                successCB:(res) => {
```

```jsx
                    console.log(res);
                    this.props.history.push("messageDetail",Object.assign(this.state.item,{chatId:res.result.chatId}))
                }
            });
        }
    render() {
        var productImgs = this.state.item.productImgs && this.state.item.productImgs.split(",")||[];
        return (
            <div>
                <Title title="详情" that={this}/>
                <div className="cm-mlr-018" onClick={()=>this.closeBox()}>
                    <div className="cm-flex cm-ai-c cm-border-bottom-ddd cm-ptb-018 cm-fs-020">
                        <img src={this.state.item.publishUserAvatar} alt="" className="cm-img-10 cm-mr-018"/>
                        <div>
                            <div className="cm-fw-bold cm-c-333 cm-fs-028">{this.state.item.publishUserName}</div>
                            <div className="cm-c-999 cm-mt-018 cm-fs-024">发布于 {getTime(this.state.item.publishTime)}</div>
                        </div>
                    </div>
                    <div>
                    <div className="cm-ptb-018 cm-fs-026">
                        <div className="cm-c-main">￥{this.state.item.productPrice}</div>
                        <div className="cm-c-333 cm-ptb-018">{this.state.item.productDesc}</div>
                    </div>
                    <div>
                        {productImgs.map((img,index)=>{
                            return(
                                <img src={img} key={index} alt="" className="cm-img-banner"/>
                            )
                        })}
                    </div>
                    </div>
```

```jsx
                        <div className="cm-ptb-018">
                            <div className="detail-line">
                                <span className="cm-mr-018">点赞</span><span>{this.state.praiseList.length}</span>
                            </div>
                            <div className="cm-flex cm-ai-c cm-ptb-018">
                                <img src={praiseHeart} alt="" className="cm-img-04 cm-mr-018"/>
                                <div className="cm-flex cm-ai-c cm-flex-wrap">
                                    {this.state.praiseList.length==0?
                                        <div>暂无点赞</div>:
                                        this.state.praiseList.map((item, index) => {
                                            return (
                                                <div className="cm-flex cm-ai-c cm-mr-018" key={index}>
                                                    <img src={item.userAvatar} className="cm-img-04 cm-border-radius-half"/>
                                                </div>
                                            )
                                        })
                                    }
                                </div>
                            </div>
                        </div>
                        <div className="cm-ptb-018">
                            <div className="detail-line"><span
                                className="cm-mr-018">留言</span><span>{this.state.commentList.length}</span></div>
                            <div className="cm-flex cm-ai-c cm-ptb-018">
                                <img src={msg} alt="" className="cm-img-04 cm-mr-018 cm-as-fs cm-mt-018"/>
                                {this.state.commentList.length==0?<div>暂无留言</div>:
                                    <div className="cm-w-full">
                                        {this.state.commentList.map((item, index) => {
                                            return (
                                                <div key={index} className="cm-ptb-018 cm-border-bottom-eee">
```

```jsx
                            <div className="cm-flex cm-jc-sb cm-ai-c">
                                <div className="cm-flex cm-ai-c">
                                    <img src={item.fromUserAvatar} alt="" className="cm-img-04 cm-border-radius-half cm-mr-018"/>
                                    <div className="cm-c-333 cm-fw-bold cm-fs-026">{item.fromUserName}</div>
                                </div>
                                <div className="cm-c-999 cm-fs-024">{getTime(item.createTime)}</div>
                            </div>
                            <div className="cm-c-666 cm-fs-height detail-replay cm-fs-026" onClick={(e)=>this.showBox(e,item)}>
                                {item.type!=1? "回复@"+item.toUserName+": ":null}{item.content}
                            </div>
                        </div>
                        )
                    })}
                </div>
            }
        </div>
    </div>
    {this.state.isShowComment?<div className="cm-bottom-position">
        <textarea ref="textarea" className="detail-textarea" name="" id="" cols="30" rows="1" placeholder=" 看对就留言，问问更多细节 "></textarea>
        <div className="cm-btn-main" onClick={()=>this.comment()}>发送</div>
    </div>:null}
    <div className="detail-fill"></div>
    <div className="cm-flex cm-ai-c cm-jc-sb detail-bottom cm-jc-sb">
        <div className="cm-flex cm-ai-c">
            <div className="cm-flex cm-ai-c cm-mr-018"><span className="cm-mr-018">点赞</span><img src={this.state.
```

```jsx
                                        isPraise == 1?alreadyPraise:praise} alt="" className="cm-img-04" onClick={()=>this.praise()}/></div>
                                        <div className="cm-flex cm-ai-c"><span className="cm-mr-018">留言</span>
                                            <img src={comment} alt="" className="cm-img-04" onClick={(e)=>this.showBox(e)}/>
                                        </div>
                                    </div>
                                    <span className="cm-btn-main" onClick={()=>this.goChat()}>我想要</span>
                                </div>
                            </div>

                </div>
            )
        }
    }
```

第 11 章 支付模块开发

【本章导读】
◎ 开发订单页面
◎ 开发订单详情页面

11.1 订单页面开发

1. 页面入口

（1）点击一级页面【首页】，选择商品，进入商品详情页面。

（2）点击【我想要】，进入聊天详情页面。

（3）点击【立即购买】，进入订单页面。

2. 页面示例

效果如图 11-1 所示。

3. 页面结构

本页面可分为 4 个部分。

第 1 部分：标题部分，可使用公共组件 Title。

第 2 部分：商品信息部分，自由编写。

第 3 部分：收货信息部分，可使用公共组件 List+ Item。

第 4 部分：底部确认部分，自由编写。

图 11-1

4. 页面数据

（1）点击标题栏返回图标，如果选择了地址，调取下单接口 placeOrder。

（2）点击【确定】，调取下单接口 placeOrder，在回调中调取支付宝接口唤起支付。

5. 编写代码

（1）创建模块及文件。

在 pages 目录下新建 payment 目录，在 payment 目录下新建 order.js 文件。

（2）配置路由。

在 router.js 文件中配置订单页面路由，代码如下。

```
import Order from './pages/payment/order';
<Route exact  path="/order"  component={Order}></Route>
```

（3）order.js 文件的代码如下。

11.2　订单详情页面开发

图 11-2

1. 页面入口

（1）点击一级页面【首页】，选择商品，进入商品详情页面。

（2）点击【我想要】，进入聊天详情页面。

（3）点击【立即购买】，进入订单页面。

（4）点击返回图标，进入订单详情页面。

2. 页面示例

效果如图 11-2 所示。

3. 页面结构

本页面可分为 6 个部分。

第 1 部分：标题部分，可使用公共组件 Title。

第 2 部分：金额与付款状态部分，自由编写。

第 3 部分：倒计时提醒部分，自由编写。

第 4 部分：商品信息部分，自由编写。

第 5 部分：订单信息（买家昵称、订单编号、交易时间）部分，自由编写。

第 6 部分：【关闭交易】和【我要付款】部分，自由编写。

4. 页面数据

(1)页面初始化时调取订单详情接口 getOrderDetails。

(2)点击【关闭交易】，调取取消交易接口 cancelOrder 接口。

(3)点击【我要付款】，调取支付宝接口使用支付功能。

5. 编写代码

(1)创建模块及文件。

在 src/pages/message 目录下新建 orderDetail.js 文件用来编写订单详情页内容，处理业务逻辑。

(2)配置路由。

在 router.js 文件中配置消息详情页路由。

```
import OrderDetail from './pages/payment/orderDetail';
<Route exact   path="/orderDetail"   component={OrderDetail}></Route>
```

(3)orderDetail.js 文件的代码如下。

```
import React from 'react';
import {Title} from '../../share';
import {goNext,getData,getTimeFormat} from '../../utils';
import Config from '../../envConfig';
export default class OrderDetail extends React.Component {
    constructor(props) {
        super(props);
        // 获取从聊天列表页面和商品详情页传过来的参数
        this.item = this.props.history.location.state || {};
        // 获取本地缓存的个人信息
        this.user = sessionStorage.getItem("userInfo")?JSON.parse(sessionStorage.getItem("userInfo")):{};
        this.state = {
            orderInfo:{}// 初始化订单信息
        }
    }
    // 获取订单详情
    componentDidMount(){
        this.getOrderDetail();
    }
```

```js
        // 如果是付款状态，直接调支付接口
        order(status){
            if(status === 1){
                window.location.href = Config.serverUrl+"alipay?orderId="+this.item.orderId+"&token="+this.user.token+"&addressId="+this.state.orderInfo.orderId;
            }
        }
        // 取消订单
        close(){
            getData({
                method:'post',
                url:'cancelOrder',
                data:{orderId:this.item.orderId||49},
                successCB:(res) => {
                    goNext(this,'index');
                }
            })
        }
        // 获取订单详情
        getOrderDetail(){
            getData({
                method:'get',
                url:'getOrderDetails',
                data:{orderId:this.item.orderId||49},
                successCB:(res) => {
                    this.setState({
                        orderInfo:res.result
                    })
                }
            })
        }
        // 按钮文本
        changeBtnText(status){
            switch (status){
                case 1:
                    return " 我要付款 ";
                case 2:
                    return " 付款中 ";
                case 3:
                    return " 已付款 ";
```

```
            case 4:
                return "付款失败";
            default:
                return "付款异常";
        }
    }
    // 付款状态
    changeStatus(status){
        switch (status){
            case 1:
                return "待付款";
            case 2:
                return "付款中";
            case 3:
                return "已付款";
            case 4:
                return "付款失败";
            default:
                return "付款异常";
        }
    }
    // 渲染视图
    render() {
        var productImgs = this.item.productImgs && this.item.productImgs.split(",");
        if(this.state.orderInfo.expiredTime>0){
            var date = parseInt(this.state.orderInfo.expiredTime/60/60/24);
            var hour = parseInt(this.state.orderInfo.expiredTime/60/60);
            var min = this.state.orderInfo.expiredTime%60;
            if(hour<10){
                hour = "0"+hour;
            }
            if(min<10){
                min = "0"+min;
            }
        }
        return <div className="cm-bc-white">
            <Title title=" 订单信息 "
                that={this}
```

```jsx
                        style={{background:"#fff"}}
                        onLeftClick={()=>goNext(this,"index")}
                />
                    <div className="cm-c-999 cm-pt-02 cm-tx-c">
                        <span className="cm-fs-040 cm-fw-bold">{this.state.orderInfo.orderAmount}</span>
                        <span>元</span>
                        <div className="cm-fs-024">{this.changeStatus(this.state.orderInfo.payStatus)}</div>
                    </div>
                    {
                        this.state.orderInfo.expiredTime>0? <div className="cm-tx-c cm-mtb-04 cm-fs-024">
                            <span className="cm-c-main">{date}</span>天
                            <span className="cm-c-main">{hour}</span>时
                            <span className="cm-c-main">{min}</span>分
后, 如果您未付款, 订单将自动关闭
                        </div>:null
                    }
                    <div className="cm-p-02 cm-border-bottom-eee cm-border-top-eee">
                        <div className="cm-mtb-01 cm-fs-024 cm-c-666"><span className="cm-fs-028 cm-c-333 cm-fw-bold">{this.state.orderInfo.consigneeName}</span><span>{this.state.orderInfo.consigneeMobile}</span></div>
                        <div className="cm-fs-024 cm-c-666">{this.state.orderInfo.consigneeAddress}</div>
                    </div>
                    <div className="cm-flex cm-jc-sb cm-bc-white cm-border-bottom-eee cm-p-02 cm-ai-c">
                        <div className="cm-flex cm-ai-c">
                            {productImgs && productImgs[0]?
                                <img src={productImgs[0]} alt="" className="cm-img-06 cm-mr-02"/>:
                                <div className="cm-img-06 cm-mr-02 cm-c-999 cm-fs-026 cm-border-ddd cm-flex cm-ai-c cm-tx-c">该商品已下架</div>
                            }
                            <div className="cm-c-333 cm-fs-026 cm-fs-ellipsis" style={{width:'220px'}}>{this.state.orderInfo.productDesc||"老家种的橙子, 甘甜又可口, 快来尝尝吧。"}</div>
                        </div>
```

```jsx
                <div className="cm-c-main">联系卖家</div>
            </div>
        <div>
            <div className="cm-mlr-02 cm-ptb-01 cm-fs-024 cm-c-999 cm-flex cm-jc-sb">
                <div>买家昵称</div>
                <div>{this.state.orderInfo.buyerUserName}</div>
            </div>
            <div className="cm-mlr-02 cm-ptb-01 cm-fs-024 cm-c-999 cm-flex cm-jc-sb">
                <div>订单编号</div>
                <div>{this.state.orderInfo.orderId}</div>
            </div>
            <div className="cm-mlr-02 cm-ptb-01 cm-fs-024 cm-c-999 cm-flex cm-jc-sb">
                <div>交易时间</div>
                <div>{getTimeFormat(this.state.orderInfo.createTime)}</div>
            </div>
        </div>
        <div className="cm-bottom-position cm-jc-sb">
            <div className="cm-mr-02 cm-ptb-02 cm-w-full cm-tx-c cm-bc-666 cm-c-white cm-border-radius-10"
                onClick={() => this.close()}>关闭交易</div>
            <div className={(this.state.orderInfo.payStatus === 1?"cm-bc-main":"cm-bc-999") +" cm-ptb-02 cm-w-full cm-tx-c cm-c-white cm-border-radius-10"}
                onClick={() => this.order(this.state.orderInfo.payStatus)}>{this.changeBtnText(this.state.orderInfo.payStatus)}</div>
        </div>
    </div>
    }
}
```

第 12 章 消息模块开发

【本章导读】

◎ 开发消息列表页面

◎ 开发消息详情页面

12.1 消息列表页面开发

图 12-1

1. 页面入口

点击一级页面【消息】,进入消息列表页面。

2. 页面示例

效果如图 12-1 所示。

3. 页面结构

本页面可分为 3 个部分。

第 1 部分:标题部分,可使用公共组件 Title。

第 2 部分:聊天记录列表部分,自由编写。

第 3 部分:底部 tab 切换部分,可使用公共组件 TabBottom。

4. 页面数据

页面初始化时获取消息列表接口 getChatList。

5. 编写代码

(1) 创建模块及文件。

在 src/pages 目录下新建 message 目录,在 message 目录下新建 messageList.js 文件用来编写消息列表页面内容,处理业务逻辑;新建 message.css 文件来编写样式。

（2）配置路由。

在 router.js 文件中配置消息列表页面路由，代码如下。

```
import MessageList from './pages/message/messageList';
<Route exact  path="/message"  component={MessageList}></Route>
```

（3）messageList.js 文件的代码如下。

```
import React from 'react';
import './message.css';
import {Title,TabBottom} from '../../share';
import {getData, goNext,getTime} from '../../utils';
export default class MessageList extends React.Component {
    constructor(props) {
        super(props);
        this.state = {
            messageList:[]// 初始化聊天列表
        }
    }
    componentDidMount(){
        this.getChatList();
    }
    // 进入聊天详情页
    goDetail(item) {
        goNext(this, 'messageDetail',item);
    }
    // 获取聊天列表内容
    getChatList(){
        getData({
            method:'post',
            url:'getChatList',
            data:{},
            successCB:(res) => {
                this.setState({
                    messageList:res.result.list
                })
            }
        })
    }
```

```jsx
        render() {
            return <div>
                <Title title=" 消息 " isHome={true}/>
                <div className="cm-mlr-018">
                    {this.state.messageList.length>0?this.state.messageList.map((item,index)=>{
                        var productImgs = item.productImgs && item.productImgs.split(",")||[];
                        var lastChatContent = item.lastChatContent && item.lastChatContent.slice(0,35);
                        return(
                        <div className="cm-flex cm-ai-c" key={index} onClick={()=>this.goDetail(item)}>
                            <img src={item.anotherUserAvatar} alt="" className="cm-img-08 cm-mr-018"/>
                            <div className="cm-flex cm-ai-c cm-jc-sb cm-w-full cm-border-bottom-ddd cm-ptb-018">
                                <div className="cm-flex-1">
                                    <div className="cm-c-333 cm-fw-bold cm-fs-028">{item.anotherUserName}</div>
                                    <div className="cm-c-666 cm-fs-015 cm-mt-018 cm-fs-026">{lastChatContent}</div>
                                    <div className="cm-c-999 cm-fs-015 cm-mt-018 cm-fs-024">{getTime(item.updateTime)}</div>
                                </div>
                                {productImgs[0]?
                                <img src={productImgs[0]} alt="" className="cm-img-12"/>:
                                <div className="cm-img-12 cm-c-999 cm-fs-026 cm-border-ddd cm-flex cm-ai-c cm-tx-c"> 该商品已下架 </div>
                                }
                            </div>
                        </div>
                        )
                    }):<div className="cm-tx-c cm-mt-080 cm-c-666"> 暂无消息 </div>}
                </div>
                <TabBottom history={this.props.history} activeNum={2}/>
            </div>
        }
    }
```

（4）message.css 文件的代码如下。

```css
.msg-detail-textarea{
    border:1px solid #ddd;
    border-radius:0.1rem;
    flex:1;
    padding:0.15rem;
    margin-right:0.18rem;
}
.detail-chat-height{
    border-radius:0.1rem;
    min-height:0.4rem;
    line-height:0.4rem;
    padding:0.1rem;
}
.detail-fixed{
    position:fixed;
    left:0;
    right:0;
    background:#fff;
    z-index:99;
}
.detail-fixed-height{
    height:1.5rem;
}
.detail-triangle{
    border-left:0.08rem solid #ddd;
    border-right:0.08rem solid transparent;
    border-top:0.08rem solid transparent;
    border-bottom:0.08rem solid transparent;
    height:0.04rem;
    margin-top:0.2rem;
}
.detail-triangle-right{
    border-left:0.08rem solid transparent;
    border-right:0.08rem solid #fff;
    border-top:0.08rem solid transparent;
    border-bottom:0.08rem solid transparent;
    height:0.04rem;
    margin-top:0.2rem;
```

```
}
.detail-ml-30{
    margin-left:0.6rem;
}
.detail-mr-30{
    margin-right:0.6rem;
}
.am-list-body{
    background-color:#f5f5f9!important;
}
```

12.2 消息详情页面开发

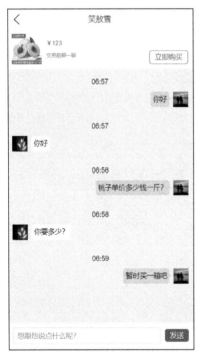

图 12-2

1. 页面入口

（1）点击一级页面【消息】，进入消息列表页面。

（2）点击某行消息进入该消息详情页面。

2. 页面示例

效果如图 12-2 所示。

3. 页面结构

本页面可分为 4 个部分。

第 1 部分：标题部分，可使用公共组件 Title。

第 2 部分：商品信息部分，自由编写。

第 3 部分：聊天记录部分，因为要进行下拉刷新，所以使用 antd-mobile 中的组件 ListView+PullToRefresh。

第 4 部分：底部发送部分，自由编写。

4. 页面数据

（1）获取聊天详情列表接口 getChatDetailList。

（2）点击【发送】，调取 websocket 接口，发送与接收聊天消息。

5. 编写代码

（1）创建模块及文件。

在 src/pages/message 目录下新建 messageDetail.js 文件用来编写消息详情页内

容,处理业务逻辑。

(2)配置路由。

在 router.js 文件中配置消息详情页面路由,代码如下。

```
import MessageDetail from './pages/message/messageDetail';
<Route exact path="/messageDetail" component={MessageDetail}></Route>
```

(3) messageDetail.js 文件的代码如下。

```
import React from 'react';
import './message.css';
import {Title} from '../../share';
import { PullToRefresh, ListView} from 'antd-mobile';
import ReactDOM from 'react-dom';
import {getData,getTimeFormat, getTimeHour,autoTextarea} from '../../utils';
let self;
export default class MessageDetail extends React.Component {
    constructor(props) {
        super(props);
        // 获取从聊天列表页面和商品详情页面传过来的参数
        this.item = this.props.history.location.state || {};
        // 获取本地缓存的个人信息
        this.user = localStorage.getItem("userInfo")?JSON.parse(localStorage.getItem("userInfo")):{};
        // 初始化聊天记录数据源
        const dataSource = new ListView.DataSource({
            rowHasChanged:(row1, row2) => row1 !== row2,
        });
        self = this;
        this.state = {
            dataSource,
            refreshing:true,
            hasMore:true,
            isLoading:true,
            height:document.documentElement.clientHeight,
            useBodyScroll:false,
            pageNum:1,// 定义初始化请求序号
```

```
            pageSize:10,// 定义初始化请求数量
            value:null,
            isShow:false
        }
        this.initData = [];
    }

    componentDidMount() {
        this.getChatDetailList();
        var chatList = this.refs.chatList;
        chatList.addEventListener("touchstart",()=>{
            this.setState({
                isShow:false
            })
        });
        this.socket = new WebSocket("ws://49.232.24.206:9001/secondary/socket?" + this.user.userId);
        // 心跳检测
        this.heartCheck = {
            timeout:60000, // 心跳间隔时间,单位为 ms
            timeoutObj:null,
            serverTimeoutObj:null,
            reset:function() {
                clearTimeout(this.timeoutObj);
                clearTimeout(this.serverTimeoutObj);
                return this;
            },
            start:function() {
                var that = this;
                this.timeoutObj = setTimeout(function() {
                    console.log(this);
                    // 这里发送一个心跳,后端收到后,返回一个心跳消息
                    //onmessage 获取到返回的心跳消息就说明连接正常
                    self.socket.send("heartbeat...【" + self.user.userId + "】");
                    that.serverTimeoutObj = setTimeout(function()
{ // 如果超过一定时间还没重置,说明后端主动断开了
                        self.socket.close(); // 如果 onclose 会执行 reconnect,我们执行 ws.close() 就行了。如果直接执行 reconnect 会触发 onclose,导致重连两次
                    }, that.timeout)
```

```
            }, this.timeout)
        }
    }
    this.socket.onopen = ()=>{
        this.heartCheck.reset().start();   // 心跳检测重置
        console.log("Socket 已打开");
    };
    this.socket.onmessage =  (msg)=> {
        this.heartCheck.reset().start();   // 心跳检测重置
        if (msg.data.indexOf("{") != -1) {
            console.log(msg);
            var obj = JSON.parse(msg.data);
            this.initData.push({content:obj.content,type:'received',time:getTimeHour(new Date().getTime())});
            this.setState({
                dataSource:this.state.dataSource.cloneWithRows(this.initData),
            })
        }
    };
    this.socket.onerror = function (err) {
        console.log(err);
    };
}
onRefresh = () => {
    if(!this.state.hasMore){
        return;
    }
    this.setState({ refreshing:true, isLoading:true });
    // simulate initial Ajax
    setTimeout(() => {
        this.getChatDetailList();
    }, 600);
};

onLeftClick() {
    this.props.history.goBack();
}
openChat(e){
    e.stopPropagation();
    this.setState({
```

```
                isShow:true
        },function () {
            autoTextarea();
        })
    }
    getChatDetailList() {
        getData({
            method:'post',
            url:'getChatDetailList',
            data:{
                chatId:this.item.chatId,
                pageNum:this.state.pageNum,
                pageSize:this.state.pageSize,
            },
            successCB:(res) => {
                this.state.pageNum++;
                var list = res.result.list;
                list.map((item)=>{
                    if(item.userId != this.user.userId){
                        item.type="received";
                    }
                    var nowTime = (new Date()).getTime();
                    var tomorrow = nowTime - 12*60*60*1000;
                    if(item.createTime<tomorrow){
                        item.time = getTimeFormat(item.createTime);
                    }else {
                        item.time = getTimeHour(item.createTime)
                    }
                });
                console.log(list);
                list = list.reverse();
                this.initData = list.concat(this.initData);// 已有的
数组拼接上拉加载更多的数组
                console.log(this.initData);
                console.dir(ReactDOM.findDOMNode(this.lv));
                const hei = this.state.height - ReactDOM.findDOMNode
(this.lv).offsetTop-50;
                let total = res.result.total;// 商品列表总数
                if(total == 0){
                    // this.setState({
                    //     statusText:" 暂无数据 "// 总数为 0 时表示没有数据
```

```
                        // })
                    }else if(total == this.initData.length){
                        this.setState({
                            dataSource:this.state.dataSource.cloneWithRows
(this.initData),
                            height:hei,
                            refreshing:false,
                            isLoading:false,
                            hasMore:false,
                            //statusText:"加载完成"// 拼接的数组长度等于总数
时表示加载完成
                        })
                    }else {
                        this.setState({
                            dataSource:this.state.dataSource.cloneWithRows
(this.initData),
                            isLoading:false,
                            refreshing:false,
                        });
                    }
                }
            })
    }
    sendMessage() {
        this.setState({
            isShow:false
        });
        var value = this.refs.textarea.value;
        if (!value) {
            return;
        }
        this.time = new Date().getTime();
        this.initData.push({content:value,time:getTimeHour(this.
time)});
        this.setState({
            dataSource:this.state.dataSource.cloneWithRows(this.
initData),
        })
        var token = localStorage.getItem("token");
        var data = {
            "toUserId":this.item.anotherUserId||this.item.
```

```jsx
                publishUserId,
                "toUserName":this.item.anotherUserName||this.item.publishUserName,
                "toUserAvatar":this.item.anotherUserAvatar||this.item.publishUserAvatar,
                "chatId":this.item.chatId,
                "token":token,
                "content":value,
            };
            if (this.socket.readyState === 1) {
                this.socket.send(JSON.stringify(data));
            } else {
                //do something
            }
        }

        changeState(e) {
            this.setState({
                value:e.target.value
            })
        }
        render() {
            var productImgs = this.item.productImgs && this.item.productImgs.split(",");
            const row = (item, sectionID, rowID) => {
                return (
                    <div key={rowID} className="cm-p-018">
                        {item.type == "received"?
                            <div className="cm-flex-column">
                                <div className="cm-tx-c cm-mtb-018">{item.time}</div>
                                <div className="cm-flex cm-ai-c cm-tx-c cm-as-fs detail-mr-30">
                                    <img src={this.item.anotherUserAvatar||this.item.publishUserAvatar} alt="" className="cm-img-06 cm-as-fs"/>
                                    <div className="cm-flex cm-flex-1">
                                        <div className="detail-triangle-right"></div>
                                        <div className="detail-chat-height cm-pad-10 cm-bc-white cm-tx-l">{item.anotherContent||item.content}</div>
```

```jsx
                                    </div>
                                </div>
                            </div>
                            :
                            <div className="cm-flex-column">
                                <div className="cm-tx-c cm-mtb-018">{item.time}</div>
                                <div className="cm-flex cm-ai-c cm-tx-c cm-as-fe detail-ml-30">
                                    <div className="cm-flex cm-flex-1">
                                        <div className=" detail-chat-height cm-pad-10 cm-bc-ddd cm-tx-l">{item.content}</div>
                                        <div className="detail-triangle"></div>
                                    </div>
                                    <img src={this.user.userAvatar} alt="" className="cm-img-06 cm-as-fs"/>
                                </div>
                            </div>
                    }
                </div>
            );
        };
        return <div>
            <Title title={this.item.anotherUserName||this.item.publishUserName}
                onLeftClick={this.onLeftClick.bind(this)}
                style={{background:"#fff"}}
            />
            <div className="detail-fixed-height" onClick={()=>this.close()}>
                <div className="cm-flex cm-jc-sb cm-p-018 detail-fixed">
                    <div className="cm-flex cm-ai-c">
                        {productImgs && productImgs[0]?
                            <img src={productImgs[0]} alt="" className="cm-img-12 cm-mr-018"/>:
                            <div className="cm-img-12 cm-mr-018 cm-c-999 cm-fs-026 cm-border-ddd cm-flex cm-ai-c cm-tx-c">该商品已下架</div>
                        }
                        <div>
                            <div className="cm-c-main cm-fs-024">￥123</div>
```

```jsx
                            <div className="cm-c-999 cm-fs-020 cm-mt-018">
交易前聊一聊</div>
                        </div>
                    </div>
                    {productImgs && productImgs[0]?
                        <div className="cm-btn-border-main cm-as-fe">立即购
买</div>:
                        <div className="cm-btn-border-999 cm-as-fe">立
即购买</div>
                    }
                </div>
            </div>
            <div onClick={()=>this.close()} ref="chatList">
                <ListView
                    key={this.state.useBodyScroll ? '0' :'1'}
                    ref={el => this.lv = el}
                    dataSource={this.state.dataSource}
                    renderRow={row}
                    useBodyScroll={this.state.useBodyScroll}
                    style={this.state.useBodyScroll ? {} :{
                        height:this.state.height
                    }}
                    pullToRefresh={<PullToRefresh
                        refreshing={this.state.refreshing}
                        onRefresh={this.onRefresh}
                    />}
                    pageSize={this.state.pageSize}
                />
            </div>
            {this.state.isShow?
            <div className="cm-bottom-position">
                <textarea onChange={(e) => this.changeState(e)}
ref="textarea" className="msg-detail-textarea" name=""
                    id="" cols="30" rows="2" placeholder="想
跟他说点什么呢?"></textarea>
                <span className="cm-btn-main cm-as-fe"
                    onClick={() => this.sendMessage()}>发送</span>
            </div>:<div className="detail-height">
                <div className="cm-bottom-position">
                    <input type="text" placeholder="想跟他说点什么
呢?" className="msg-detail-textarea" onClick={(e) => this.openChat(e)}/>
```

```
                    <span className="cm-btn-main">发送</span>
                </div>
            </div>}
        </div>
    }
}
```

第 13 章 个人中心模块开发

【本章导读】

◎ 开发个人中心页面

◎ 开发编辑个人信息页面

◎ 开发我发布的、卖出的、买到的相关页面

◎ 开发修改/忘记密码页面

◎ 开发修改手机号页面

◎ 开发管理收货地址页面

图 13-1

13.1 个人中心页面开发

1. 页面入口

点击一级页面【我的】，进入个人中心页面。

2. 页面示例

效果如图 13-1 所示。

3. 页面结构

本页面可分为 5 个部分。

第 1 部分：标题部分，可使用公共组件 Title。

第 2 部分：头像与昵称部分，自由编写，点击头像，可进入编辑个人信息页面。

第 3 部分：个人商品管理、修改密码与修改手机号部分，可使用公共组件 List。

第 4 部分：【安全退出】部分，可使用公共组件 Button。

第 5 部分：底部 tab 切换部分，可使用公共组件 TabBottom。

4. 页面数据

（1）页面初始化时获取个人中心发布的、卖出的、买到的商品数量与金额接口 getProductNum。

(2)点击【安全退出】,调取退出登录接口 logout。

5. 编写代码

(1)创建模块及文件。

在 src/pages 目录下新建 personal 目录,在 message 目录下新建 personal.js 文件用来编写个人中心页内容,处理业务逻辑,新建 personal.css 文件来编写样式。

(2)配置路由。

在 router.js 文件中配置个人中心页路由,代码如下。

```
import Personal from './pages/personal/personal';
<Route exact  path="/personal"  component={Personal}/>
```

(3)personal.js 文件的代码如下。

```
import React from 'react';
import './personal.css';
import sale from '../../images/personal/sale.png';
import buy from '../../images/personal/buy.png';
import publish from "../../images/personal/publish.png";
import modify from '../../images/personal/modify.png';
import {Title, List,Button,TabBottom} from '../../share';
import {getData, goNext} from '../../utils';
const Item = List.Item;
export default class Personal extends React.Component {
    constructor(props) {
        super(props);
        this.state = {
            info:[
                {leftTitle:"我发布的",leftIcon:[publish],total:0,link:"myPublish"},
                {leftTitle:"我卖出的",leftIcon:[sale],total:0,link:"mySale"},
                {leftTitle:"我买到的",leftIcon:[buy],total:0,link:"myPurchase"},
                {leftTitle:"修改密码",leftIcon:[modify],link:"modifyPwd"},
                {leftTitle:"修改手机号",leftIcon:[modify],link:"verifyOldMobile"}
```

```
            ],// 定义个人中心列表初始化
            sales:0// 我卖出的商品金额初始化为 0
        }
    }
    componentDidMount(){
        this.getProductNum();
    }
    // 退出登录
    goLogin(){
        getData({
            method:'post',
            url:'logout',
            data:{},
            successCB:(res) => {
                // 清除本地缓存
                localStorage.removeItem("userInfo");
                localStorage.removeItem("token");
                // 进入登录页面
                goNext(this, 'login');
            }
        })
    }
    // 获取个人中心发布的、卖出的、买到的商品数量与金额
    getProductNum(){
        getData({
            method:'get',
            url:'getProductNum',
            successCB:(res) => {
                if(res.code == 0){
                    this.state.info[0].total = res.result.publish.num;
                    this.state.info[1].total = res.result.sale.num;
                    this.state.info[2].total = res.result.purchase.num;
                    this.setState({
                        sales:res.result.sale.money
                    })
                }
            }
        })
    }
    render() {
        var user = JSON.parse(localStorage.getItem("userInfo"))||{};
```

```jsx
        return <div>
            <Title title=" 个人中心 " isHome={true}/>
            <div className="cm-mlr-018">
                <div className="cm-mtb-018 cm-ptb-018 cm-flex cm-ai-c" onClick={() => goNext(this, 'modifyInfo')}>
                    <img src={user.userAvatar} alt="" className="cm-img-10 cm-border-radius-half"/>
                    <div className="cm-ml-018">
                        <div className="cm-c-333 cm-fw-bold cm-fs-028">{user.userName}</div>
                        <div className="cm-c-999 cm-mt-018 cm-fs-022">当前卖出累计金额 <span className="cm-c-main cm-fs-026">￥{this.state.sales}</span>
                        </div>
                    </div>
                </div>
                <div className="cm-space-line"></div>
                {this.state.info.map((item,index)=>{
                    return(
                    <List key={index} onClick={()=>goNext(this,item.link,item)}>
                        <Item
                            leftIcon={item.leftIcon}
                            leftTitle={item.leftTitle}
                            rightTitle={item.total}
                        />
                    </List>
                    )
                })}
                <Button className="cm-mt-080" onClick={()=>this.goLogin()}>安全退出 </Button>
                <TabBottom history={this.props.history} activeNum={3}/>
            </div>
        </div>
    }
}
```

（4）personal.css 文件的代码如下。

```css
.personal-input{
```

```css
    width:1.6rem;
    height:1.6rem;
    opacity:0;
}
.personal-again,.personal-input-box{
    width:1.6rem;
    border-radius:50%;
    background:#ffffff;
    border:1px solid #ddd;
    margin:1rem auto auto auto;
    height:1.6rem;
    color:#ddd;
}
.personal-input-box:before,.personal-input-box:after{
    content:"";
    height:0.5rem;
    width:0.05rem;
    background:#ddd;;
    position:absolute;
    top:50%;
    left:50%;
    transform:translate(-50%, -50%)
}
.personal-input-box:after{
    height:0.05rem;
    width:0.5rem;
}
.personal-btn{
    background:#cc0000;
    padding:0.15rem 0.2rem;
    color:#fff;
    border-radius:0.1rem;
}
```

13.2 编辑个人信息页面开发

1. 页面入口

（1）点击一级页面【我的】，进入个人中心页面。

（2）点击头像，进入编辑个人信息页面。

2. 页面示例

效果如图 13-2 所示。

3. 页面结构

本页面可分为 4 个部分。

第 1 部分：标题部分，使用公共组件 Title。

第 2 部分：头像部分，自由编写，点击头像，可选择图片。

第 3 部分：昵称部分，可使用公共组件 List+Input。

第 4 部分：【确认修改】部分，可使用公共组件 Button。

4. 页面数据

点击【确认修改】，调取更新个人信息接口 updateUserInfo。

5. 编写代码

图 13-2

（1）创建模块及文件。

在 src/pages/personal 目录下新建 modifyInfo.js 文件用来编写编辑个人信息内容，处理业务逻辑。

（2）配置路由。

在 router.js 文件中配置编辑个人信息路由，代码如下。

```
import ModifyInfo from './pages/personal/modifyInfo';
<Route exact  path="/modifyInfo"   component={ModifyInfo}></Route>
```

（3）modifyInfo.js 文件的代码如下。

```
import React from 'react';
import './personal.css';
import {Title,List,Input,Button} from '../../share';
import {getData, goNext,checkParam} from '../../utils';
const Item = List.Item;
export default class ModifyInfo extends React.Component {
    constructor(props) {
        super(props);
```

```
            // 获取本地缓存的个人信息
            let user = localStorage.getItem("userInfo")?JSON.parse
(localStorage.getItem("userInfo")):{};
            this.state = {
                userAvatar:[user.userAvatar],
                userName:user.userName,
            }
        }
        componentDidMount(){
            this.uploadImg();
        }
        onLeftClick(){
            this.props.history.goBack();
        }
        change(state){
            this.setState(state)
        }
        // 修改个人信息
        modify(){
            var arr = [
                {value:this.state.userAvatar,msg:"请选择头像"},
                {value:this.state.userName,msg:"请输入昵称"},
            ]
            checkParam(arr,()=> {
                console.log(this.state.userAvatar);
                var token = localStorage.getItem("token");
                let param = new FormData(); // 创建form对象
                param.append('userAvatar',this.state.userAvatar[0]);//
通过append向form对象添加数据
                param.append('userName',this.state.userName);
                getData({
                    method:'post',
                    url:'updateUserInfo',
                    headers:{
                        'Content-Type':'multipart/form-data',
                        'token':token
                    },
                    data:param,
                    successCB:(res) => {
                        // 修改成功之后，替换本地缓存的头像和昵称，并进入个人中心
页面
```

```
                    var user = JSON.parse(localStorage.getItem
("userInfo"));
                    user.userAvatar = this.state.userAvatar[0];
                    user.userName = this.state.userName;
                    localStorage.setItem("userInfo",JSON.stringify
(user));
                    goNext(this,'personal');
                }
            })
        })
    }
    // 选择个人头像
    getAvatar(event){
        var files = event.target.files;
        this.setState({
            userAvatar:files
        });
        if(window.FileReader) {
            var file = files[0];
            var fr = new FileReader();
            fr.readAsDataURL(file);    // 将图片作为URL读出
            fr.onloadend = (e)=> {
                this.refs.avatar.style.backgroundImage  = "url("+e.
target.result+")";
                this.refs.avatar.style.backgroundSize = "100%";
            };
        }
        this.refs.avatar.setAttribute("class","personal-again");
    }
    // 反显个人头像
    uploadImg(e) {
        let user = localStorage.getItem("userInfo");
        user = localStorage.getItem("userInfo")?JSON.parse(localStorage.
getItem("userInfo")):{};
        this.refs.avatar.style.backgroundImage  = "url("+user.
userAvatar+")";
        this.refs.avatar.style.backgroundSize = "100%";
        this.refs.avatar.setAttribute("class","personal-again");
    }
    render() {
        return<div>
```

```
                <Title  onLeftClick={()=>this.onLeftClick()} title="编
辑个人信息"/>
                <div className="cm-mt-036">
                    <div className="personal-input-box" ref="avatar">
                        <input type="file" className="personal-input"
onChange={(e)=>this.getAvatar(e)}/>
                    </div>
                </div>
                <div className="cm-m-018 cm-mt-018">
                    <List>
                        <Input
                            label="昵称"
                            style = {{textAlign:"right"}}
                            maxLength={16}
                            placeholder="请输入昵称"
                            defaultValue={this.state.userName}
                            onChange={(val)=>this.change({userName:val})}
                            className="cm-c-999"
                        />
                    </List>
                    <Button type="fill" className="cm-mt-080" onClick=
{() => this.modify()}>确认修改</Button>
                </div>
            </div>
        }
    }
```

13.3 我发布的商品列表页面开发

1. 页面入口

（1）点击一级页面【我的】，进入个人中心页面。

（2）点击【我发布的】，进入我发布的商品列表页面。

（3）点击某行可进入我发布的商品详情页面。

2. 页面示例

效果如图 13-3 所示。

图 13-3

3. 页面结构

本页面可分为 3 个部分。

第 1 部分：标题部分，可使用公共组件 Title。

第 2 部分：我发布的商品数量统计部分，自由编写。

第 3 部分：我发布的商品列表部分，可使用公共组件 Card。点击【编辑】，可修改发布的商品信息；点击【下架】，可删除发布的商品信息。

4. 页面数据

页面初始化时调取我发布的商品列表接口 getMyProductList。

5. 编写代码

（1）创建模块及文件。

在 src/pages/personal 目录下新建 myPublish.js 文件用来编写我发布的商品信息内容，处理业务逻辑。因为我发布的、我卖出的、我买到的商品列表页面都是卡片形式，所以考虑将这种卡片列表封装成一个公共组件。在 src/pages/personal 目录下新建 productList.js 文件作为公共组件。

（2）配置路由。

在 router.js 文件中配置我发布的商品路由，公共组件不需要配置路由，代码如下。

```
import MyPublish from './pages/personal/myPublish';
<Route exact  path="/myPublish"  component={MyPublish}></Route>
```

（3）productList.js 文件的代码如下。

```
import React from 'react';
import {Card} from '../../share';
import {getTimeFormat} from '../../utils';
export default class ProductList extends React.Component {
    constructor(props) {
        super(props);
    }
    render() {
        const {dataList,TopArea,ButtonArea,NoData} = this.props;
        return <div>
                <TopArea num={dataList.length}/>
                <div className="cm-plr-018">
```

```
                        {dataList.length>0?dataList.map((item,index)=>{
                            var productImgs = item.productImgs && item.productImgs.split(",");
                            var desc = item.productDesc && item.productDesc.slice(0,28);
                            return(
                                <Card key={index}>
                                    <div className="cm-flex cm-jc-sb cm-ai-c cm-plr-009 cm-ptb-018">
                                        <div className="cm-flex cm-ai-c">
                                            <img src={productImgs[0]} alt="" className="cm-img-14"/>
                                            <div className="cm-ml-018 cm-flex-1">
                                                <div className="cm-c-333 cm-fs-026">{desc}</div>
                                                <div className="cm-c-main cm-fs-030 cm-mt-018">￥{item.productPrice}</div>
                                                <div className="cm-c-999 cm-pt-018 cm-fs-022">{getTimeFormat(item.createTime)}</div>
                                            </div>
                                        </div>
                                    </div>
                                    <ButtonArea item={item}/>
                                </Card>
                            )
                        }):
                        <NoData/>
                    }
                </div>
            </div>
        }
    }
```

（4）myPublish.js 文件的代码如下。

```
import React from 'react';
import {Title} from '../../share';
import publish from "../../images/personal/publish.png";
import {getData, goNext} from '../../utils';
```

```
import ProductList from './productList';
let self;
export default class MyPublish extends React.Component {
    constructor(props) {
        super(props);
        self = this;
        this.state = {
            dataList:[]// 初始化我发布的商品列表
        }
    }
    componentDidMount(){
        this.getMyProductList(1);
    }
    // 获取我发布的商品列表
    getMyProductList(type){
        getData({
            method:'post',
            url:'getMyProductList',
            data:{
                type:type
            },
            successCB:(res) => {
                this.setState({
                    dataList:res.result.list
                })
            }
        })
    }
    render() {
        return <div>
            <Title title=" 我发布的 " that={this} style={{background:"#fff"}}/>
            <ProductList
                TopArea={TopArea}
                NoData={NoData}
                ButtonArea={ButtonArea}
                dataList = {this.state.dataList}
            />
        </div>
    }
}
```

```jsx
class ButtonArea extends React.Component{
    constructor(props){
        super(props);
    }
    // 编辑已发布的信息，进入商品信息编辑页面
    modifyPublish(e,item){
        e.stopPropagation();
        goNext(this,"modifyProduct",item);
    }
    // 删除已发布的商品信息
    delProduct(e,item){
        e.stopPropagation();
        getData({
            method:'post',
            url:'delProduct',
            data:{
                productId:item.id
            },
            successCB:(res) => {
                // 删除之后重新调取我发布的商品列表接口，刷新列表页
                self.getMyProductList(1);
            }
        })
    }
    render(){
        return  <div className="cm-flex cm-ai-c cm-ptb-018 cm-border-top-eee cm-jc-fe">
            <div className="cm-btn-border-333 cm-flex cm-ai-c cm-mr-018" onClick={(e)=>this.modifyPublish(e)}>
                <span className="cm-fs-028">下架</span>
            </div>
            <div className="cm-btn-border-333 cm-flex cm-ai-c cm-mr-018" onClick={(e)=>this.delProduct(e)}>
                <span className="cm-fs-028">下架</span>
            </div>
        </div>
    }
}
class TopArea extends React.Component{
    constructor(props){
        super(props);
```

```
                console.log(this);
            }
            render(){
                return <div className="cm-ptb-018 cm-p-018 cm-bc-white">
                    <div className="cm-flex cm-ai-c">
                        <img src={publish} alt="" className="cm-img-04 cm-border-radius-half"/>
                        <div className="cm-c-999 cm-ml-018 cm-fs-026">我发布的商品共<span className="cm-c-main cm-fs-028"> {this.props.num} </span>件</div>
                    </div>
                </div>
            }
        }
        class NoData extends React.Component{
            constructor(props){
                super(props);
            }
            render(){
                return  <div className="cm-c-666 cm-tx-c cm-mt-080">
                    <div>你还没有发布任何宝贝哦！</div>
                    <div className="cm-mt-036" onClick={()=>goNext(self,'publish')}>
                        <span className="cm-btn-main-higher">发布宝贝</span>
                    </div>
                </div>
            }
        }
```

13.4 我发布的商品信息编辑页面开发

1. 页面入口

（1）点击一级页面【我的】，进入个人中心页面。

（2）点击【我发布的】，进入我发布的商品列表页面。

（3）点击【编辑】，可进入我发布的商品信息编辑页面。

2. 页面示例

效果如图13-4所示。

图 13-4

3. 页面结构

本页面可分为 6 个部分。

第 1 部分：标题部分，可使用公共组件 Title。

第 2 部分：商品描述部分，自由编写。

第 3 部分：选择图片部分，可使用 antd-mobile 中的 ImagePicker 组件。

第 4 部分：文本框部分，可使用公共组件 List + Input + Select。

第 5 部分：【确认修改】部分，可使用公共组件 Button。

第 6 部分：底部 tab 切换部分，可使用公共组件 TabBottom。

4. 页面数据

（1）从我发布的商品列表页面进入，携带参数地址信息。

（2）点击【确认修改】，调取更新商品接口 updateProduct。

5. 编写代码

（1）创建模块及文件。

在 src/pages/personal 目录下新建 modifyProduct.js 文件用来编写我发布的商品信息内容，处理业务逻辑。

（2）配置路由。

在 router.js 文件中配置商品信息修改路由，代码如下。

```
import MyPublish from './pages/personal/modifyProduct;
<Route exact  path="/ modifyProduct "  component={ModifyProduct}>
</Route>
```

（3）modifyProduct.js 文件的代码如下。

```
import React from 'react';
import {ImagePicker} from 'antd-mobile';
import {Select,Title,List,Input,Button} from '../../share';
```

```jsx
import {getData,goNext,checkParam,isDefine} from '../../utils';
export default class ModifyProduct extends React.Component {
    constructor(props) {
        super(props);
        this.state = {
            productImgs:[],
            typeList:[],
            productType:'',
            productPrice:0,
            productDesc:'',
        };
        // 获取从我发布的商品列表页面传过来的参数，并分别赋值
        this.item = this.props.history.location.state ||{};
        this.state.productId = this.item.id;
        this.state.productDesc = this.item.productDesc;
        this.state.productPrice = this.item.productPrice;
        this.state.productType = this.item.productTypeId;
        this.state.productTypeName = this.item.productTypeName;
        this.state.productImgs = [];
        var imgList = this.item.productImgs.split(',');
        imgList.map((item)=>{
            this.state.productImgs.push({url:item});
        })
    }
    componentDidMount(){
        // 检查登录状态
        getData({
            method:'post',
            url:'checkLoginValid',
        });
        this.getTypeList();
    }
    // 选择商品分类
    changeType(item){
        this.setState({
            productType :item.id
        })
    }
    // 选择商品图片
    onChange = (productImgs, type, index) => {
        this.setState({
```

```
            productImgs,
        });
    }
    // 添加商品描述
    changeDes(event){
        this.setState({
            productDesc:event.target.value
        })
    }
    // 金额正则校验
    changePrice(event){
        event.target.value = event.target.value.replace(/[^\d.]/g,"");
        // 清除数字和"."以外的任意字符
        event.target.value = event.target.value.replace(/\.{2,}/g,".");
        // 只保留第一个".", 清除多余连续的"."
        event.target.value = event.target.value.replace(".","$#$").replace(/\./g,"").replace("$#$",".");
        // 将第一个"."用特殊符号替换，其他"."都转换为空，再将特殊符号用"."
        // 替换回来，目的是保证清除非连续的"."
        event.target.value = event.target.value.replace(/^(\-)*(\d+)\.(\d\d).*$/,'$1$2.$3');
        // 只能输入两位小数
        if(event.target.value.indexOf(".")< 0 && event.target.value
!=""){
            // 以上已经过滤，此处控制的是，如果没有小数点，首位不能为类似于 01、
02 的金额
            event.target.value= parseFloat(event.target.value);
        }
        // 通过 setState 改变金额
        this.setState({
            productPrice:event.target.value
        })
    }
    // 获取商品分类列表信息
    getTypeList(){
        if(isDefine(localStorage.getItem("typeList"))){
            this.setState({
                typeList:JSON.parse(localStorage.getItem("typeList"))
            })
        }
    }
```

```js
submit(){
    var arr = [
        {value:this.state.productId, msg:"请上传商品编号"},
        {value:this.state.productDesc, msg:"请输入商品描述"},
        {value:this.state.productImgs[0], msg:"请添加图片"},
        {value:this.state.productPrice, msg:"请输入商品价格"},
        {value:this.state.productType, msg:"请选择分类"},
    ];
    checkParam(arr,()=>{
        var token = localStorage.getItem("token");
        let param = new FormData(); // 创建form对象
        for (var i = 0; i < this.state.productImgs.length; i++){
            param.append('productImgs',this.state.productImgs[i].file);// 通过append向form对象添加数据
        }
        param.append('productId',this.state.productId);
        param.append('productDesc',this.state.productDesc);
        param.append('productPrice',this.state.productPrice);
        param.append('productTypeId',this.state.productType);
        getData({
            method:'post',
            url:'updateProduct',
            headers:{
                'Content-Type':'multipart/form-data',
                'token':token
            },
            data:param,
            successCB:(res) => {
                goNext(this,"index");
            }
        })
    })
}
change(state){
    this.setState(state)
}
render(){
    const { productImgs } = this.state;
    return(
        <div>
            <Title  title="编辑商品信息" that={this}/>
```

```jsx
                    <div className="cm-mlr-018 cm-mt-018">
                        <textarea name="" id="" cols="30"
                            onChange={(e)=>this.changeDes(e)}
                            defaultValue={this.state.productDesc}
                            maxLength="30"
                            className="cm-w-full cm-border-ddd cm-p-018" rows="10" placeholder="描述商品转手原因、入手渠道和使用感受"></textarea>
                        <ImagePicker
                            files={productImgs}
                            onChange={this.onChange}
                            onImageClick={(index, fs) => console.log(index, fs)}
                            selectable={productImgs.length < 9}
                        />
                    </div>
                    <div className="cm-mlr-018">
                        <List>
                            <Input
                                label="价格"
                                style = {{textAlign:"right"}}
                                placeholder="请输入商品价格"
                                defaultValue={this.state.productPrice}
                                onChange={(val)=>this.change({productPrice:val})}
                                maxLength={11}
                            />
                        </List>
                        <List>
                            <Select
                                options={this.state.typeList}
                                babel="分类"
                                onChange={this.changeType.bind(this)}
                                activeOne={{type:this.state.productType, name:this.state.productTypeName}}
                            />
                        </List>
                    </div>
                    <Button className="cm-mt-080" onClick={()=>this.submit()}>确认修改</Button>
                </div>
            )
```

        }
    }

13.5 我卖出的商品列表页面开发

1. 页面入口

（1）点击一级页面【我的】，进入个人中心页面。

（2）点击【我卖出的】，进入我卖出的商品列表页面。

2. 页面示例

效果如图 13-5 所示。

3. 页面结构

本页面可分为 3 个部分。

第 1 部分：标题部分，可使用公共组件 Title。

第 2 部分：我卖出的商品数量统计部分，自由编写。

第 3 部分：我卖出的商品列表部分，可使用公共组件 Card。点击【查看】，可查看我卖出的订单详情；点击【删除】，可删除我卖出的商品信息。

图 13-5

4. 页面数据

页面初始化时调取我卖出的商品列表接口 getMyProductList。

5. 编写代码

（1）创建模块及文件。

在 src/pages/personal 目录下新建 mySale.js 文件用来编写我卖出的商品信息内容，处理业务逻辑。

（2）配置路由。

在 router.js 文件中配置我卖出的商品路由，代码如下。

```
import MyPurchase from './pages/personal/mySale';
<Route exact  path="/mySale"   component={MySale}></Route>
```

（3）mySale.js 文件的代码如下。

```js
import React from 'react';
import {Title} from '../../share';
import sale from "../../images/personal/sale.png";
import ProductList from './productList';
import {getData, goNext} from '../../utils';
let self;
export default class MySale extends React.Component {
    constructor(props) {
        super(props);
        self = this;
        this.state = {
            dataList:[],
            value:null
        }
    }
    componentDidMount(){
        this.getMyProductList(2);
    }
    getMyProductList(type){
        getData({
            method:'post',
            url:'getMyProductList',
            data:{
                type:type
            },
            successCB:(res) => {
                this.setState({
                    dataList:res.result.list
                });
            }
        })
    }
    render() {
        return <div>
            <Title title=" 我卖出的 " that={this} style={{background:"#fff"}}/>
            <ProductList
            TopArea={TopArea}
```

```
                NoData={NoData}
                ButtonArea={ButtonArea}
                dataList = {this.state.dataList}
                 />
            </div>
        }
    }
    class ButtonArea extends React.Component{
        constructor(props){
            super(props);
            console.log(props);
        }
        goDetail(e){
            e.stopPropagation();
            goNext(self,"indexDetail",this.props.item);
        }
        delProduct(e){
            e.stopPropagation();
            getData({
                method:'post',
                url:'delProduct',
                data:{
                    productId:this.props.item.id
                },
                successCB:(res) => {
                    self.getMyProductList(2);
                }
            })
        }
        render(){
            return   <div className="cm-flex cm-ai-c cm-ptb-018 cm-border-top-eee cm-jc-fe">
                <div className="cm-btn-border-333 cm-flex cm-ai-c cm-mr-018" onClick={(e)=>this.goDetail(e)}>
                    <span className="cm-fs-028">查看</span>
                </div>
                <div className="cm-btn-border-333 cm-flex cm-ai-c cm-mr-018" onClick={(e)=>this.delProduct(e)}>
                    <span className="cm-fs-028">删除</span>
                </div>
            </div>
```

```
    }
}
class TopArea extends React.Component{
    constructor(props){
        super(props);
    }
    render(){
        return <div className="cm-ptb-018 cm-p-018 cm-bc-white">
            <div className="cm-flex cm-ai-c">
                <img src={sale} alt="" className="cm-img-04 cm-border-radius-half"/>
                <div className="cm-c-999 cm-ml-018 cm-fs-026">我卖出的宝贝共 <span className="cm-c-main cm-fs-028"> {this.props.num} </span>件 </div>
            </div>
        </div>
    }
}
class NoData extends React.Component{
    constructor(props){
        super(props);
    }
    render(){
        return  <div className="cm-c-666 cm-tx-c cm-mt-080">
            <div> 你还没有卖出任何宝贝哦！</div>
            <div className="cm-mt-036" onClick={()=>goNext(self,'index')}>
                <span className="cm-btn-main-higher"> 去首页瞧瞧！</span>
            </div>
        </div>
    }
}
```

13.6　我买到的商品列表页面开发

1.页面入口

（1）点击一级页面【我的】，进入个人中心页面；

（2）点击【我买到的】，进入我买到的商品列表页面。

2. 页面示例

效果如图 13-6 所示。

3. 页面结构

本页面可分为 3 个部分。

第 1 部分：标题部分，可使用公共组件 Title。

第 2 部分：我买到的商品数量统计部分，自由编写。

第 3 部分：我买到的商品列表部分，可使用公共组件 Card。点击【查看】，可查看我买到的订单详情页；点击【删除】，可删除买到的商品信息。

图 13-6

4. 页面数据

初始化页面时调取我买到的商品列表接口 getMyProductList。

5. 编写代码

（1）创建模块及文件。

在 src/pages/personal 目录下新建 myPurchase.js 文件用来编写我买到的商品信息内容，处理业务逻辑。

（2）配置路由。

在 router.js 文件中配置我买到的商品路由，代码如下。

```
import MyPurchase from './pages/personal/myPurchase';
<Route exact  path="/myPurchase"  component={MyPurchase}></Route>
```

（3）myPurchase.js 文件的代码如下。

```
import React from 'react';
import {Title} from '../../share';
import buy from "../../images/personal/buy.png";
import ProductList from './productList';
import {getData, goNext} from '../../utils';
let self;
export default class MySale extends React.Component {
    constructor(props) {
```

```
        super(props);
        self = this;
        this.state = {
            dataList:[]
        }
    }
    componentDidMount(){
        this.getMyProductList(3);
    }
    getMyProductList(type){
        getData({
            method:'post',
            url:'getMyProductList',
            data:{
                type:type
            },
            successCB:(res) => {
                this.setState({
                    dataList:res.result.list
                });
            }
        })
    }
    render() {
        return <div>
            <Title title=" 我买到的 " that={this} style={{background:"#fff"}}/>
            <ProductList
                TopArea={TopArea}
                NoData={NoData}
                ButtonArea={ButtonArea}
                dataList = {this.state.dataList}
            />
        </div>
    }
}
class TopArea extends React.Component{
    constructor(props){
        super(props);
        console.log(this);
    }
```

```jsx
        render(){
            return <div className="cm-ptb-018 cm-p-018 cm-bc-white">
                <div className="cm-flex cm-ai-c">
                    <img src={buy} alt="" className="cm-img-04 cm-border-radius-half"/>
                    <div className="cm-c-999 cm-ml-018 cm-fs-026">我买到的商品共<span className="cm-c-main cm-fs-028"> {this.props.num} </span>件</div>
                </div>
            </div>
        }
    }
    class ButtonArea extends React.Component{
        constructor(props){
            super(props);

        }
        goDetail(e){
            e.stopPropagation();
            goNext(self,"indexDetail",this.props.item);
        }
        delProduct(e){
            e.stopPropagation();
            getData({
                method:'post',
                url:'delProduct',
                data:{
                    productId:this.props.item.id
                },
                successCB:(res) => {
                    self.getMyProductList(3);
                }
            })
        }
        render(){
            return  <div className="cm-flex cm-ai-c cm-ptb-018 cm-border-top-eee cm-jc-fe">
                <div className="cm-btn-border-333 cm-flex cm-ai-c cm-mr-018" onClick={(e)=>this.goDetail(e)}>
                    <span className="cm-fs-028">查看</span>
                </div>
                <div className="cm-btn-border-333 cm-flex cm-ai-c cm-mr-
```

```
018"  onClick={(e)=>this.delProduct(e)}>
                    <span className="cm-fs-028"> 删除 </span>
                </div>
            </div>
        }
    }
    class NoData extends React.Component{
        constructor(props){
            super(props);
        }
        render(){
            return  <div className="cm-c-666 cm-tx-c cm-mt-080">
                <div> 你还没有买到任何宝贝哦！ </div>
                <div className="cm-mt-036" onClick={()=>goNext(self,'index')}>
                    <span className="cm-btn-main-higher"> 去首页瞧瞧！ </span>
                </div>
            </div>
        }
    }
```

13.7　修改密码页面开发

图 13-7

1. 页面入口

（1）点击一级页面【我的】，进入个人中心页面。

（2）点击【修改密码】，进入修改密码页面。

2. 页面示例

效果如图 13-7 所示。

3. 页面结构

本页面可分为 3 个部分。

第 1 部分：标题部分，可使用公共组件 Title。

第 2 部分：输入密码部分，可使用公共组件 List + Input。

第 3 部分：【确认修改】部分，修改成功后返回个人中心页面，可使用公共组件 Button。

4. 页面数据

点击【确认修改】，调取更新密码接口 updatePwd。

5. 编写代码

（1）创建模块及文件。

在 src/pages/personal 目录下新建 modifyPwd.js 文件用来编写修改密码内容，处理业务逻辑。

（2）配置路由。

在 router.js 文件中配置修改密码的路由，代码如下。

```
import ModifyPwd from './pages/personal/modifyPwd';
<Route exact  path="/modifyPwd"  component={ModifyPwd}></Route>
```

（3）modifyPwd.js 文件的代码如下。

```
import React from 'react';
import {Title,List,Input,Button} from '../../share';
import {getData,goNext,checkParam} from '../../utils';
import {Toast} from 'antd-mobile';
export default class ModifyPwd extends React.Component {
    constructor(props) {
        super(props);
        this.state = {
            password:'',
            newPassword:"",
            confirmPwd:'',
        }
    }
    change(state){
        this.setState(state)
    }
    submit(){
        var arr = [
            {value:this.state.password,msg:"请输入旧密码"},
            {value:this.state.newPassword,msg:"请输入新密码"},
            {value:this.state.confirmPwd,msg:"请输入确认密码"},
            {value:this.state.newPassword == this.state.confirmPwd,msg:"密码与确认密码不一致"},
        ]
        checkParam(arr,()=> {
```

```jsx
            getData({
                method:'post',
                url:'updatePwd',
                data:{
                    "oldPwd":this.state.password,
                    "newPwd":this.state.newPassword,
                    "confirmPwd":this.state.confirmPwd
                },
                successCB:(res) => {
                    Toast.success(res.message);
                    goNext(this,"personal")
                }
            })
        })
    }
    render() {
        return<div>
            <Title  that={this} title=" 修改密码 "/>
            <div className="cm-mlr-018 cm-mt-018">
                <List>
                    <Input
                        type="password"
                        placeholder=" 请输入旧密码 "
                        onChange={(val)=>this.change({password:val})}
                        maxLength={16}
                    />
                </List>
                <List>
                    <Input
                        type="password"
                        placeholder=" 请输入新密码 "
                        onChange={(val)=>this.change({newPassword:val})}
                        maxLength={16}
                    />
                </List>
                <List>
                    <Input
                        type="password"
                        placeholder=" 请输入确认密码 "
                        value={this.state.confirmPwd}
                        onChange={(val)=>this.change({confirmPwd:val})}
```

```
                          maxLength={16}
                        />
                      </List>
                      <Button type="fill" className="cm-mt-080" onClick={()
=> this.submit()}>确认修改 </Button>
                    </div>
                </div>
            )
    }
}
```

13.8 忘记密码页面开发

1. 页面入口

（1）点击一级页面【我的】，未登录时进入登录页面。

（2）点击【忘记密码】，进入忘记密码页面。

2. 页面示例

效果如图 13-8 所示。

3. 页面结构

本页面可分为 3 个部分。

第 1 部分：标题部分，可使用公共组件 Title。

第 2 部分：输入手机号、验证码和密码部分，可使用公共组件 List + Input。

第 3 部分：【完成】部分，设置密码成功后进入登录页面，可使用公共组件 Button。

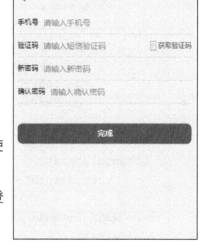

图 13-8

4. 页面数据

（1）点击【获取验证码】，调取获取验证码接口 getVerifyCode。

（2）点击【完成】，调取忘记密码接口 forgotPwd。

（3）点击【完成】，调取绑定新手机接口 bindNewMobile，成功后将本地 localStorage 的手机号替换。

5. 编写代码

（1）创建模块及文件。

在 src/pages/personal 目录下新建 forgotPwd.js 文件，作为忘记密码组件。

（2）配置路由。

在 router.js 文件中配置忘记密码组件的路由，代码如下。

```
import ForgotPwd from './pages/personal/forgotPwd';
<Route exact path="/forgotPwd" component={ForgotPwd}></Route>
```

（3）forgotPwd.js 文件的代码如下。

```
import React from 'react';
import {Title,List,Input,Button} from '../../share';
import {getData, goNext,isDefine,checkParam} from '../../utils';
import {Toast} from 'antd-mobile';
import verifyCode from '../../images/register/verify-code.png';
export default class ForgotPwd extends React.Component {
    constructor(props) {
        super(props);
        this.state = {
            mobile:"",
            password:'',
            verifyCode:"",
            getCode:" 获取验证码 ",
            confirmPwd:'',
        }
    }
    change(state){
        this.setState(state)
    }
    // 调取忘记密码接口
    submit(){
        // 非空校验数组对象
        var arr = [
            {value:this.state.mobile,msg:" 请输入手机号 "},
            {value:this.state.verifyCode,msg:" 请输入短信验证码 "},
            {value:this.state.password,msg:" 请输入新密码 "},
            {value:this.state.confirmPwd,msg:" 请输入确认密码 "},
            {value:this.state.password == this.state.confirmPwd,msg:" 密码与确认密码不一致 "},
        ];
        checkParam(arr,()=> {
```

```js
        getData({
            method:'post',
            url:'forgotPwd',
            data:{
                "mobile":this.state.mobile,
                "verifyCode":this.state.verifyCode,
                "newPwd":this.state.password,
                "confirmPwd":this.state.confirmPwd
            },
            successCB:(res) => {
                // 密码重置成功后进入登录页面
                goNext(this,"login")
            }
        })
    })
}
// 验证手机号
getVerify(){
    if(!isDefine(this.state.mobile)){
        Toast.info(" 请输入手机号 ");
        return;
    }
    getData({
        method:'post',
        url:'getVerifyCode',
        data:{
            mobile:this.state.mobile,
            type:2
        },
        successCB:(res)=> {
            this.setState({
                getCode:60
            },()=>{
                this.timer = setInterval(()=>{
                    // 验证码过了 60 秒，消除定时器
                    if(this.state.getCode == 1){
                        clearInterval(this.timer);
                        this.setState({
                            getCode:" 获取验证码 "
                        })
                    }else {
```

```
                            // 验证码 60 秒倒计时
                            this.setState({
                                getCode:--this.state.getCode
                            })
                        }
                    },1000)
                })
            }
        }
        renderRight(){
            return(
                <div className="cm-flex cm-ai-c"><img src={verifyCode} alt="" className="cm-img-04"/><span className="cm-c-main" onClick={()=>this.getVerify()}> {this.state.getCode == " 获取验证码 "?" 获取验证码 ":this.state.getCode+" 秒重新获取 "}</span></div>
            )
        }
        render() {
            return<div>
                <Title  onLeftClick={()=>this.onLeftClick()} title=" 忘记密码 "/>
                <div className="cm-mlr-018 cm-mt-018">
                    <List>
                        <Input
                            label=" 手机号 "
                            maxLength={11}
                            placeholder=" 请输入手机号 "
                            defaultValue={this.state.mobile}
                            onChange={(val)=>this.change({mobile:val})}
                            className="cm-c-999"
                        />
                    </List>
                    <List>
                        <Input
                            label=" 验证码 "
                            placeholder=" 请输入短信验证码 "
                            defaultValue={this.state.verifyCode}
                            onChange={(val)=>this.change({verifyCode:val})}
                            renderRight = {()=>this.renderRight()}
                        />
```

```
                </List>
                <List>
                    <Input
                        label=" 新密码 "
                        type="password"
                        placeholder=" 请输入新密码 "
                        onChange={(val)=>this.change({password:val})}
                        maxLength={16}
                    />
                </List>
                <List>
                    <Input
                        label=" 确认密码 "
                        type="password"
                        placeholder=" 请输入确认密码 "
                        value={this.state.confirmPwd}
                        onChange={(val)=>this.change({confirmPwd:val})}
                        maxLength={16}
                    />
                </List>
                <Button type="fill" className="cm-mt-080" onClick=
{() => this.submit()}> 完成 </Button>
            </div>
        </div>
    }
}
```

13.9 修改手机号页面开发

1. 页面入口

（1）点击一级页面【我的】，进入个人中心页面。

（2）点击【修改手机号】，进入验证原手机页面和绑定新手机页面。

2. 页面示例

效果如图 13-9 和图 13-10 所示。

3. 页面结构

本页面可分为 3 个部分。

第 1 部分：标题部分，使用公共组件 Title。

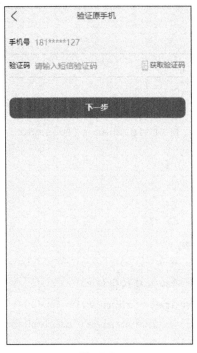

图 13-9　　　　　　　　　　　　图 13-10

第 2 部分：手机号和验证码部分，使用公共组件 List 组件 + Input 组件。

第 3 部分：点击【完成】按钮，手机号修改成功后进入个人中心页面。

注意：修改手机号有两个流程，首先需要验证原手机号，然后绑定新手机号。两个页面布局类似，可以使用公共组件 ModifyMobile。

4．页面数据

（1）点击【获取验证码】，调取获取验证码接口 getVerifyCode。

（2）点击【下一步】，调取验证原手机号接口 verifyOldMobile。

（2）点击【完成】，调取绑定新手机接口 bindNewMobile，成功后将本地 localStorage 的手机号替换。

5．编写代码

（1）创建模块及文件。

在 src/pages/personal 目录下新建 modifyMobile.js 文件，作为修改手机号的公共组件。

在 src/pages/personal 目录下新建 verifyOldMobile.js 文件，作为验证原手机的组件。

在 src/pages/personal 目录下新建 bindNewMobile.js 文件，作为绑定新手机的

组件。

（2）配置路由。

在 router.js 文件中配置验证原手机组件和绑定新手机组件的路由，代码如下。

```
import VerifyOldMobile from './pages/personal/verifyOldMobile';
import BindNewMobile from './pages/personal/bindNewMobile';
<Route exact  path="/verifyOldMobile"  component={VerifyOldMobile}></Route>
<Route exact  path="/bindNewMobile"  component={BindNewMobile}></Route>
```

（3）modifyMobile.js 文件的代码如下。

```
import React from 'react';
import {List,Input,Button} from '../../share';
import {getData,isDefine} from '../../utils';
import {Toast} from 'antd-mobile';
import verifyCode from '../../images/register/verify-code.png';
export default class ModifyMobile extends React.Component {
    constructor(props) {
        super(props);
        this.state = {
            getCode:'获取验证码',
            mobile:props.mobile?props.mobile:"",// 验证原手机号的时候手机号直接从父组件传递过来
            verifyCode:"",// 初始化验证码
        };
        this.type = props.type;
    }
    change(state){
        this.setState(state)
    }
    // 验证手机号
    getVerify(){
        if(!isDefine(this.state.mobile)){
            Toast.info(" 请输入手机号 ");
            return;
        }
```

```
            getData({
                method:'post',
                url:'getVerifyCode',
                data:{
                    mobile:this.state.mobile,
                    type:this.type
                },
                successCB:(res)=> {
                    this.setState({
                        getCode:60
                    },()=>{
                        this.timer = setInterval(()=>{
                            // 验证码过了60秒，消除定时器
                            if(this.state.getCode == 1){
                                clearInterval(this.timer);
                                this.setState({
                                    getCode:" 获取验证码 "
                                })
                            }else {
                                // 验证码60秒倒计时
                                this.setState({
                                    getCode:--this.state.getCode
                                })
                            }
                        },1000)
                    })
                }
            })
        }
        renderRight(){
            return(
                <div className="cm-flex cm-ai-c"><img src={verifyCode}
alt="" className="cm-img-04"/><span className="cm-c-main" onClick={()
=>this.getVerify()}> {this.state.getCode == " 获取验证码 "?" 获取验证码 ":this.
state.getCode+" 秒重新获取 "}</span></div>
            )
        }
        // 点击下一步或者完成的回调
        submit(){
            if(this.props.submit){
                this.props.submit({mobile:this.state.mobile,verifyCode:
```

```
this.state.verifyCode});
            }
        }
        render() {
            return <div>
                    <div className="cm-mlr-018 cm-mt-018">
                        <List>
                            <Input
                                label=" 手机号 "
                                maxLength={11}
                                placeholder=" 请输入手机号 "
                                defaultValue={this.state.mobile}
                                onChange={(val)=>this.change({mobile:val})}
                                className="cm-c-999"
                            />
                        </List>
                        <List>
                            <Input
                                label=" 验证码 "
                                placeholder=" 请输入短信验证码 "
                                maxLength={6}
                                defaultValue={this.state.verifyCode}
                                onChange={(val)=>this.change({verifyCode:val})}
                                renderRight = {()=>this.renderRight()}
                            />
                        </List>
                        <Button type="fill" className="cm-mt-080" onClick={() => this.submit()}>{this.props.buttonText}</Button>
                    </div>
                </div>
        }
    }
```

（4）verifyOldMobile.js 文件的代码如下。

```
import React from 'react';
import {Title} from '../../share';
import {getData, goNext,checkParam} from '../../utils';
import ModifyMobile from './modifyMobile';
```

```jsx
export default class VerifyNewMobile extends React.Component {
    constructor(props) {
        super(props);
    }
    // 验证原手机
    modify(opt){
        console.log(opt);
        var arr = [
            {value:opt.mobile,msg:"请输入手机号"},
            {value:opt.verifyCode,msg:"请输入验证码"},
        ];
        checkParam(arr,()=> {
            var param = {
                mobile:opt.mobile,
                verifyCode:opt.verifyCode
            };
            getData({
                method:'post',
                url:'verifyOldMobile',
                data:param,
                successCB:(res) => {
                    // 原手机号验证成功之后清除验证码的定时器，进入绑定新手机页面
                    clearInterval(this.timer);
                    goNext(this,"bindNewMobile");
                }
            })
        })
    }
    render() {
        var user = localStorage.getItem("userInfo")?JSON.parse(localStorage.getItem("userInfo")):{};
        var mobile = user.mobile.slice(0,3) + "*****"+user.mobile.slice(user.mobile.length-3,user.mobile.length);
        return(
            <div>
                <Title that={this} title=" 验证原手机 "/>
                <ModifyMobile
                 mobile = {mobile}
                 buttonText = " 下一步 "
                 submit = {(opt)=>this.modify(opt)}
                 type = {3}
```

```
                />
            </div>
        )
    }
}
```

(5) bindNewMobile.js 文件的代码如下。

```
import React from 'react';
import {Title} from '../../share';
import {getData, goNext,checkParam} from '../../utils';
import ModifyMobile from './modifyMobile';
export default class BindNewMobile extends React.Component {
    constructor(props) {
        super(props);
    }
    // 绑定新手机号
    submit(opt){
        var arr = [
            {value:opt.mobile,msg:"请输入新手机号"},
            {value:opt.verifyCode,msg:"请输入验证码"},
        ]
        checkParam(arr,()=> {
            var param = {
                mobile:opt.mobile,
                verifyCode:opt.verifyCode
            };
            getData({
                method:'post',
                url:'bindNewMobile',
                data:param,
                successCB:(res) => {
                    // 绑定成功之后清除验证码的定时器，将个人信息重新缓存一
份，进入个人中心页面
                    clearInterval(this.timer);
                    var user = JSON.parse(localStorage.getItem
("userInfo"));
                    user.mobile = opt.mobile;
                    localStorage.setItem("userInfo",JSON.stringify
(user));
                    goNext(this,"personal")
```

```
                    }
                })
            })
        }
        render() {
            return(
                <div>
                    <Title  that={this} title=" 绑定新手机 "/>
                    <ModifyMobile
                        buttonText = " 完成 "
                        submit = {(opt)=>this.submit(opt)}
                        type = {4}
                    />
                </div>
            )
        }
    }
```

13.10 我的收货地址列表页面开发

图 13-11

1. 页面入口

（1）点击一级页面【我的】，进入个人中心页面；

（2）点击【我的收货地址】，进入我的收货地址列表页面。

2. 页面示例

效果如图 13-11 所示。

3. 页面结构

本页面可分为 2 个部分。

第 1 部分：标题部分，可使用公共组件 Title。

第 2 部分：我的收货地址列表部分，自由编写。如果设置了默认地址，会有默认字样；当地址列表为空时，显示"添加地址"，通过"添加地址"和标题列的"添加"，可进入添加地址页面。点击【编辑】，可对地址进行修改。

4. 页面数据

（1）从订单详情页进入，携带参数地址信息。

（2）不是从订单详情页进入，调取我的收货地址列表接口 getAddressList。

5. 编写代码

（1）创建模块及文件。

在 src/pages/personal 目录下新建 addressManage.js 文件，用来编写我的收货地址列表内容，处理业务逻辑。

（2）配置路由。

在 router.js 文件中配置我的收货地址页路由，代码如下。

```
import AddressManage from './pages/personal/addressManage';
<Route exact  path="/addressManage"  component={AddressManage}></Route>
```

（3）addressManage.js 文件的代码如下。

```
import React from 'react';
import {Title} from '../../share';
import {getData, goNext} from '../../utils';

export default class AddressManage extends React.Component {
    constructor(props) {
        super(props);
        this.state = {
            addressList:[]// 初始化列表
        };
        // 从订单页面进来获取页面来源信息。this.item 为 {from:order}
        this.item = this.props.history.location.state ||{};
    }
    componentDidMount(){
        this.getAddressList();
    }
    // 获取地址列表内容
    getAddressList(){
        getData({
            method:'post',
            url:'getAddressList',
            data:{
```

```jsx
                    pageNum:1,
                    pageSize:100
                },
                successCB:(res) => {
                    console.log(res);
                    this.setState({
                        addressList:res.result.list
                    })
                }
            })
        }
        // 如果从订单页面进来，点击列表返回订单页面
        goPage(item){
            if(this.item.from == "order"){
                goNext(this,'order',item);
            }
        }
        render() {
            return <div>
                <Title title=" 我的收货地址 " that={this} rightTitle=" 添加 " onRightClick={()=>goNext(this,'addAddress')}/>
                <div className="cm-mlr-018">
                    {this.state.addressList.length>0?this.state.addressList.map((item,index)=>{
                        var street = item.street?item.street:"";
                        return(
                            <div className={this.item.id == item.id?"cm-flex cm-ai-c cm-bc-white":"cm-flex cm-ai-c"}
                                key={index} onClick={()=>this.goPage(item)}>
                                <div className="cm-flex cm-ai-c cm-jc-sb cm-flex-1 cm-border-bottom-ddd cm-ptb-018">
                                    <div className="cm-flex-1 cm-mr-018">
                                        <div className="cm-c-333 cm-fs-028 cm-fw-bold"><span>{item.consigneeName}</span><span className="cm-c-666 cm-fs-026"> {item.consigneeMobile}</span></div>
                                        <div className="cm-c-999 cm-fs-026 cm-mt-009">{item.isDefaultAddress === 1?<span className="cm-p-006 cm-c-white" style={{background:"pink",marginRight:"0.1rem"}}>默认
```

```
</span>":""}<span>{item.province+item.city+item.district+street+item.
addressDetail}</span></div>
                                        </div>
                                        <div className="cm-btn-border-333"
onClick={(e)=>
                                        {e.stopPropagation();goNext(this,
'addAddress',item)}}
                                        >编辑</div>
                                    </div>
                                </div>
                        )
                    }):<div className="cm-tx-c cm-ptb-018 cm-border-
bottom-ddd cm-c-666" onClick={()=>goNext(this,'addAddress')}>添加地址</
div>}
                </div>
            </div>
        }
    }
```

13.11 添加/编辑收货地址页面开发

13.11.1 添加收货地址

1. 页面入口

(1) 点击一级页面【我的】,进入个人中心页面。

(2) 点击【我的收货地址】,进入我的收货地址列表页面。

(3) 初始进入我的收货地址列表页面,无任何地址信息,可点击【添加收货地址】,进入添加收货地址页面。

(4) 点击标题栏的【添加】,也可以进入添加收货地址页面。

2. 页面示例

效果如图 13-12 所示。

3. 页面结构

本页面可分为 4 个部分。

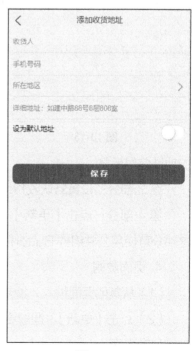

图 13-12

第 1 部分：标题部分，可使用公共组件 Title。

第 2 部分：添加收货地址信息（收货人、手机号、所在地区、详细地址）部分，可使用公共组件 List+Input，其中所在地区可使用封装的 Address 组件。

第 3 部分：设置默认收货地址部分，可使用 antd-mobile 中的 Switch 组件。

第 4 部分：【保存】部分，保存后可添加一条收货地址信息，可使用公共组件 Button。

4. 页面数据

点击【保存】，调取保存收货地址接口 saveAddress。

13.11.2　编辑收货地址

1. 页面入口

（1）点击一级页面【我的】，进入个人中心页面．

（2）点击【我的收货地址】，进入我的收货地址列表页面。

（3）点击我的收货地址列表页面的【编辑】，进入编辑收货地址页面。

2. 页面示例

效果如图 13-13 所示。

3. 页面结构

本页面可分为 4 个部分。

第 1 部分：标题部分，可使用公共组件 Title。

图 13-13

第 2 部分：编辑收货地址信息部分，自由编写。可使用公共组件 List+Input，获取从地址列表页面传过来的地址信息反显。

第 3 部分：设置默认收货地址部分，可使用 antd-mobile 中的 Switch 组件。

第 4 部分：点击【更新】，保存收货地址信息；点击【删除】，删除收货地址信息。这部分可使用公共组件 Button。

4. 页面数据

（1）从其他页面进入，携带地址参数信息。

（2）点击【更新】，调取更新地址接口 updateAddress，更新地址后，进入我的收货地址列表页面。

（3）点击【删除】，调取删除地址接口 delAddress，删除地址后，进入我的收货地址列表页面。

13.11.3 编写代码

（1）创建模块及文件。

在 src/pages/personal 目录下新建 addAddress.js 文件，用来编写添加 / 编辑收货地址内容，处理业务逻辑。

（2）配置路由。

在 router.js 文件中配置添加 / 编辑收货地址页路由，代码如下。

```
import AddAddress from './pages/personal/addAddress';
<Route exact  path="/addAddress"  component={AddAddress}></Route>
```

（3）addAddress.js 的代码如下。

```
import React from 'react';
import {Title, List, Input,Button,Address} from '../../share';
import {getData, goNext, checkParam} from '../../utils';
import {Switch} from 'antd-mobile';
const Item = List.Item;
export default class AddAddress extends React.Component {
    constructor(props) {
        super(props);
        this.state = {
            id:"",// 地址编号
            consigneeName:'',// 收货人
            consigneeMobile:'',// 收货人联系方式
            province:"",// 省份
            city:"",// 城市
            district:"",// 区县
            street:"",// 街道
            addressDetail:'',// 详细地址
            isDefaultAddress:false,// 是否为默认地址
            isShowAddress:false// 是否显示地址组件
        };
```

```
            // 作为编辑地址时其他页面携带过来的参数信息
            this.item = this.props.history.location.state ||{};
            // 如果有参数信息携带过来，重新赋值地址信息
            if(this.item.consigneeName){
                this.state.id = this.item.id;
                this.state.consigneeName = this.item.consigneeName;
                this.state.consigneeMobile = this.item.consigneeMobile;
                this.state.province = this.item.province;
                this.state.city = this.item.city;
                this.state.district = this.item.district;
                this.state.street = this.item.street;
                this.state.addressDetail = this.item.addressDetail;
                // 地址为1时转义为布尔值true，即默认地址；地址为2时为false
                this.state.isDefaultAddress = this.item.isDefaultAddress === 1?true:false;
            }
        }
        // 改变value值时重置state
        change(state) {
            this.setState(state)
        }
        // 调取删除地址接口
        delete(){
            var arr = [
                {value:this.state.id, msg:"没有收货地址编号"}
            ];
            checkParam(arr, () => {
                getData({
                    method:'post',
                    url:'delAddress',
                    data:{
                        id:this.state.id,
                    },
                    successCB:(res) => {
                        goNext(this, 'addressManage')
                    }
                })
            })
        }
        // 调取更新地址接口
        update(){
```

```
            var arr = [
                {value:this.state.id, msg:"没有收货地址编号"},
                {value:this.state.consigneeName, msg:"请输入收货人"},
                {value:this.state.consigneeMobile, msg:"请输入收货人联系方
式"},
                {value:this.state.province, msg:"请输入省份"},
                {value:this.state.city, msg:"请输入城市"},
                {value:this.state.district, msg:"请输入县区"},
                {value:this.state.addressDetail, msg:"请输入详细地址"},
            ];
            checkParam(arr, () => {
                getData({
                    method:'post',
                    url:'updateAddress',
                    data:{
                        id:this.state.id,
                        consigneeName:this.state.consigneeName,
                        consigneeMobile:this.state.consigneeMobile,
                        province:this.state.province,
                        city:this.state.city,
                        district:this.state.district,
                        street:this.state.street,
                        addressDetail:this.state.addressDetail,
                        isDefaultAddress:this.state.isDefaultAddress?1:2,
                    },
                    successCB:(res) => {
                        goNext(this, 'addressManage')
                    }
                })
            })
        }
        // 调取保存地址接口
        save() {
            var arr = [
                {value:this.state.consigneeName, msg:"请输入收货人"},
                {value:this.state.consigneeMobile, msg:"请输入收货人联系方
式"},
                {value:this.state.province, msg:"请输入省份"},
                {value:this.state.city, msg:"请输入城市"},
                {value:this.state.district, msg:"请输入县区"},
                {value:this.state.addressDetail, msg:"请输入详细地址"},
```

```
            ];
            checkParam(arr, () => {
                getData({
                    method:'post',
                    url:'saveAddress',
                    data:{
                        consigneeName:this.state.consigneeName,
                        consigneeMobile:this.state.consigneeMobile,
                        province:this.state.province,
                        city:this.state.city,
                        district:this.state.district,
                        street:this.state.street,
                        addressDetail:this.state.addressDetail,
                        isDefaultAddress:this.state.isDefaultAddress?1:2,
                    },
                    successCB:(res) => {
                        goNext(this, 'addressManage')
                    }
                })
            })
        }
        // 显示地址下拉列表框
        showAddressModel(){
            this.setState({
                isShowAddress:true
            })
        }
        // 关闭地址下拉列表框
        closeAddressModel(){
            this.setState({
                isShowAddress:false
            });
        }
        // 关闭地址下拉列表框，获取地址信息
        getAddress(address){
            this.setState({
                isShowAddress:false,
                province:address.province,
                city:address.city,
                district:address.district,
                street:address.street,
```

```jsx
            });
        }
        render() {
            return (
                <div>
                    <Title that={this} title={this.item.consigneeName?"
编辑收货地址":"添加收货地址"}/>
                    <div className="cm-mlr-018">
                        <List>
                            <Input
                                placeholder=" 收货人 "
                                defaultValue={this.state.consigneeName}
                                onChange={(val) => this.change
({consigneeName:val})}
                                style={{textAlign:'left'}}
                            />
                        </List>
                        <List>
                            <Input
                                type="number"
                                maxLength={11}
                                placeholder=" 手机号码 "
                                defaultValue={this.state.consigneeMobile}
                                onChange={(val) => this.change
({consigneeMobile:val})}
                                style={{textAlign:'left'}}
                            />
                        </List>
                        <List onClick={()=>this.showAddressModel()}>
                            <Item
                                leftTitle={
                                    this.state.province? <div>
                                        <div>{this.state.province}
</div>
                                        <div>{this.state.city}</div>
                                        <div>{this.state.district}
</div>
                                        <div>{this.state.street}
</div>
                                    </div>
                                    :<div className="cm-c-999"> 所在地区
```

```jsx
                </div>
                            }
                        />
                    </List>
                    <List>
                        <Input
                            type="text"
                            placeholder=" 详细地址：如建中路 88 号 8 层 808 室 "
                            defaultValue={this.state.addressDetail}
                            onChange={(val) => this.change
({addressDetail:val})}
                            style={{textAlign:'left'}}
                        />
                    </List>
                    <div className="cm-flex cm-ptb-018 cm-jc-sb">
                        <div>设为默认地址</div>
                        <Switch
                            checked={this.state.isDefaultAddress}
                            onChange={(val) => {
                                this.setState({
                                    isDefaultAddress:val,
                                });
                            }}
                        />
                    </div>
                    {this.item.consigneeName?<div className="cm-flex cm-jc-sb"><Button type="half" className="cm-mt-080" onClick={() => this.update()}>更新</Button>
                            <Button type="half"  className="cm-mt-080" onClick={() => this.delete()}>删除</Button>
                        </div>:<Button type="fill" className="cm-mt-080" onClick={() => this.save()}>保存</Button>}

                </div>
                {this.state.isShowAddress?<Address
                        closeModel={()=>this.closeAddressModel()}
                        getAddress={(address)=>this.getAddress(address)}
                    />:null}
            </div>)
        }
    }
```

第 14 章 前端环境部署

【本章导读】
◎ 项目部署流程

项目部署流程

前端环境部署相对比较简单，主要分为以下 4 个步骤。

1. 更改当前环境

一般项目中会有一个文件统一配置项目环境。如下所示，当 appType 为 1 时，表示本地开发环境；当 appType 为 2 时，表示测试环境；当 appType 为 3 时，表示生产环境。

```
var appType = 2;
const Config = {};
if (appType == 1) {
    // 本地开发环境
    Config.serverUrl = 'http://localhost:9001/secondary/';
    Config.serverIp = 'http://192.168.1.103:3000/';
} else if (appType == 2) {
    // 测试环境
    Config.serverUrl = 'http://49.232.24.206:9001/secondary/';
    Config.serverIp = 'http://49.232.24.206:9001/';
}else if (appType == 3){
    // 生产环境
    Config.serverUrl = 'http://49.232.24.206:9001/secondary/';
    Config.serverIp = 'http://49.232.24.206:9001/';
}else {
    Config.serverUrl = 'http://49.232.24.206:9001/secondary/';
    Config.serverIp = 'http://49.232.24.206:9001/';
```

```
}
export default Config;
```

比较完善的环境一般有5套，即DEV开发环境、SIT环境、UAT环境、准生产环境、生产环境。

DEV 开发环境：专门用于开发的服务器，配置可以比较随意。为了节约服务器资源，也可以将本地作为开发环境。为了开发调试方便，一般打开全部错误日志。

SIT 环境：System Integration Test，即系统集成测试，开发人员自己测试流程是否正确。一般只有内网才能访问。

UAT 环境：User Acceptance Test，即用户验收测试，由专门的测试人员验证，验收完成才能进入生产环境。一般也只有内网才能访问。

准生产环境：环境配置、数据库、页面展示与生产环境完全一样。为了使生产环境万无一失，一般会准备一套准生产环境，外网可以访问，但是需要加入白名单。有时候在浏览百度或者支付宝等大型应用时，会提示是否加入内部测试之类的文字，这就是部署的准生产环境。

生产环境：正式提供对外服务，一般会关闭错误报告，打开错误日志。可以理解为包含所有功能的环境，任何项目所使用的环境都以这个为基础，然后根据用户的个性化需求来进行调整或者修改。

项目开发环境没有严格的规定，很多企业一般只准备 3 套环境，即开发环境、测试环境和生产环境。

2. 打包

环境更改之后就开始进行打包，打包命令为 npm run-script build。打包之后根目录会生成一个 build 文件夹，里面的文件就是打包之后的文件。

3. 配置 nginx

步骤 1：安装依赖。

安装 gcc、pcre-devel、zlib-devel 和 openssl-devel

gcc：GNU 操作系统的编译器

pcre：正则表达式库

zlib：提供数据压缩用的函式库

openssl：开放源代码的软件库包，使用这个包来进行安全通信，避免窃听，同时确认另一端连接者的身份

-devel 的安装包一般是开发软件的包，用于编译的时候连接的库之类的文件
安装依赖命令如下：

```
yum install gcc zlib zlib-devel openssl openssl--devel pcre pcre-devel
```

步骤 2：下载安装 nginx。
① 下载 nginx 的 tar 包并解压，命令如下。

```
tar -zxvf nginx-1.11.3.tar.gz
```

② 安装（安装到 /usr/local/nginx 目录，分别执行以下 3 个命令）。

```
./configure --prefix=/usr/local/nginx
make（编译）
make install（安装）
```

步骤 3：启动 nginx。
① 进入 usr/local/nginx/sbin 目录，命令如下。

```
cd usr/local/nginx/sbin
```

② 查看 nginx 进程，命令如下。

```
ps -ef | grep nginx
```

③ 执行启动命令如下。

```
./nginx
```

步骤 4：配置单页面应用。
① 进入 usr/local/nginx/conf 目录，命令如下。

```
cd usr/local/nginx/conf
```

② 打开 nginx.conf 文件命令如下。

`vi nginx.conf`

③ 编辑 nginx.conf 文件，按 Insert 键插入字符，添加 try_files $uri $uri//index.html。

④ 保存并退出。

先按下 Esc，再输入 wq

4. 部署

在官方网站下载 FileZilla 工具，使用 FileZilla 工具将打包出来的文件上传到服务器。

（1）打开 FileZilla 工具，输入服务器 IP 和密码，端口默认为 22，点击【快速连接】。

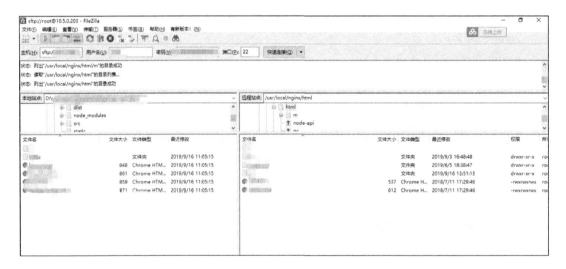

（2）左边的"本地站点"是本地项目目录，找到打包之后的文件。

（3）右边的"远程站点"是服务器目录，找到需要上传的目录地址。

（4）将左边的所有文件选中并右击，在弹出的快捷菜单中选择"上传"，将所有的文件上传到服务器。

第 15 章 前端开发总结

【本章导读】

◎ 开发思路总结

◎ 项目难点总结

15.1 开发思路总结

不管是一个完整的项目开发，还是一个需求开发，或者是一个简单的页面开发，其实开发思路都是一样的，那就是在开发前需要整体规划。首先考虑将公共样式、公共组件提炼出来。其次分析页面结构，确定哪些内容可以用第三方库快速开发。比如轮播图功能，就可以直接使用 antd-mobile 组件库开发，如果自己封装，有一定难度不说，即便封装出来，因为涉及动画，组件性能也可能会不佳。但是其他内容，比如标题、按钮、文本框等，这些结构封装比较简单，一方面，自己封装的组件会根据项目需求量身定制，有更好的兼容性；另一方面也锻炼自己封装公共组件的能力，提高编程能力。

15.2 项目难点总结

本项目难点主要有 4 个方面。

1. textarea 自动获取高度

textarea 自动获取高度指的是当输入的内容高度超过文本框高度时，自动增加文本框的高度。

解决办法如下：

```
export function autoTextarea(){
    var textarea = document.getElementsByTagName("textarea");
    for (var i = 0; i < textarea.length; i++) {
        textarea[i].style.height = '0.4rem';
        textarea[i].scrollTop = 0;  // 防抖动
        textarea[i].style.height = textarea[i].scrollHeight + 'px';
        textarea[i].addEventListener('input', function (e) {
            e.target.style.height = 'auto';
            e.target.scrollTop = 0;  // 防抖动
            if (e.target.scrollHeight < 20) {
                e.target.style.height = '20px';
            } else {
                e.target.style.height = e.target.scrollHeight + 'px';
            }
        })
    }
}
```

分析：

（1）获取所有的 textarea 标签；

（2）首先将 textarea 高度初始化为 0.4rem，然后将高度重置为元素滑动的高度；

（3）监听 input 事件，如果高度小于 20px（0.4rem*50px），则重置为 20px，否则在 input 事件的监听下，将高度重置为元素滑动的高度。

2. 上传头像 / 图片

场景为多张图片时，可使用 antd-mobile 组件库提供的 ImagePicker 组件。如果只有一张图片，可将 input 文本框的 type 属性设置为 file，图片上传到服务器时，需要使用 FileReader 方法将图片作为二进制流传输。

```
var files = event.target.files;
this.setState({
    userAvatar:files
});
if(window.FileReader) {
    var file = files[0];
    var fr = new FileReader();
    fr.onloadend = (e)=> {
        this.refs.avatar.style.backgroundImage  = "url("+e.target.
```

```
result+")";
            this.refs.avatar.style.backgroundRepeat = "no";
            this.refs.avatar.style.backgroundSize = "100%";
        };
        // 将图片作为 URL 读出
        fr.readAsDataURL(file);
    }
```

3. 上拉加载与下拉刷新

上拉加载主要通过 antd-mobile 组件库提供的 ListView 组件，与 onEndReached 方法结合使用。注意，页面滑动到底部时才触发方法 onEndReached。

上拉加载的实现思路是页面初始化的时候加载 10 条数据，当上拉到顶部时，继续加载后面的 10 条数据，以此类推。

```
<ListView
        useBodyScroll={true}
        ref={el => this.lv = el}
        dataSource={this.state.dataSource}
        renderFooter={() => (<div style={{ paddingBottom:60, textAlign:'center' }}>
            {this.state.statusText}
        </div>)}
        renderBodyComponent={() => <Body/>}
        renderRow={row}
        pageSize={4}
        scrollRenderAheadDistance={500}
        onEndReached={this.onEndReached}
        onEndReachedThreshold={10}
    />
```

下拉刷新是通过 antd-mobile 组件库提供的 PullToRefresh 组件来实现。需要注意的是，在下拉刷新时，需要实时更改滑动区域的高度。

```
<ListView
    key={this.state.useBodyScroll ? '0' :'1'}
    ref={el => this.lv = el}
    dataSource={this.state.dataSource}
```

```
        renderRow={row}
        useBodyScroll={this.state.useBodyScroll}
        style={this.state.useBodyScroll ? {} :{
              height:this.state.height
           }}
        pullToRefresh={<PullToRefresh
           refreshing={this.state.refreshing}
           onRefresh={this.onRefresh}
        />}
        pageSize={this.state.pageSize}
/>
```

4. 实时聊天

实时聊天主要使用 HTML5 提供的 WebSoket 技术，主要有以下几种方法。

（1）创建 WebSoket，方法为 new WebSoket()。

（2）打开 WebSoket，方法为 soket.onopen=function(){}。

（3）发送数据，方法为 soket.send()。

（4）关闭连接，方法为 soket.close()。

（5）WebSoket 关闭时触发，方法为 soket.onclose=function(){}。

（6）WebSoket 连接出错，方法为 soket.onerror=function(){}。

（7）监听服务端返回数据，方法为 soket.onmessage=function(){}。

```
    this.socket = new WebSocket("ws://49.232.24.206:9001/secondary/
socket?" + this.user.userId);
    // 心跳检测
    this.heartCheck = {
        timeout:60000, // 心跳间隔时间，单位为 ms
        timeoutObj:null,
        serverTimeoutObj:null,
        reset:function() {
            clearTimeout(this.timeoutObj);
            clearTimeout(this.serverTimeoutObj);
            return this;
        },
        start:function() {
            var that = this;
            this.timeoutObj = setTimeout(function() {
```

```
                console.log(this);
                // 这里发送一个心跳，后端收到后，返回一个心跳消息
                //onmessage 获取到返回的心跳就说明连接正常
                self.socket.send("heartbeat...【" + self.user.userId + "】");
                that.serverTimeoutObj = setTimeout(function() { // 如果超
过一定时间还没重置，说明后端主动断开了
                    self.socket.close(); // 如果 onclose 会执行 reconnect，
我们执行 ws.close() 就行了。如果直接执行 reconnect 会触发 onclose，导致重连两次
                }, that.timeout)
            }, this.timeout)
        }
    }
    this.socket.onopen = ()=>{
        this.heartCheck.reset().start(); // 心跳检测重置
        console.log("Socket 已打开");
    };
    this.socket.onmessage =  (msg)=> {
        this.heartCheck.reset().start(); // 心跳检测重置
        if (msg.data.indexOf("{") != -1) {
            console.log(msg);
            var obj = JSON.parse(msg.data);
            this.initData.push({content:obj.content,type:'received',time:getTimeHour(new Date().getTime())});
            this.setState({
                dataSource:this.state.dataSource.cloneWithRows(this.initData),
            })
        }
    };
    this.socket.onerror = function (err) {
        console.log(err);
    };
```

后端开发

第 16 章　Java 基础

【本章导读】
◎ Java 特点
◎ Java 语法
◎ Java 开发环境

16.1　Java 主要特点

Java 是一种简单、面向对象、分布式、编译与解释性、健壮性、安全性、可移植性、高性能、多线程和动态的编程语言。程序员可以用 Java 编写桌面应用程序、Web 应用程序、分布式系统等应用程序。

简单：Java 看起来跟 C++ 很相似，但是为了使语法简洁、容易上手，去掉了很多 C++ 里面的特征。例如，去掉了重载（overload）操作，也去掉了烦琐的多继承与指针，实现了自动垃圾回收，程序员不必为内存管理而烦恼。

面向对象：Java 吸取了 C++ 面向对象的特点，将数据封装于类中，利用类的优点，实现了程序的简洁性、便于维护性。类的封装、继承、多态等特性，使程序代码只需一次编译就可以反复利用。

分布式：Java 包含了一个支持 HTTP 和 FTP 等基于 TCP/IP 的子库。因此，Java 应用程序可凭借 URL 打开并访问网络上的对象，其访问方式与访问本地文件系统几乎完全相同。

编译与解释性：Java 是先编译成 class 文件，然后再利用虚拟机解释执行的，所以 Java 是一种先编译后解释的语言。

健壮性：Java 是一种强类型语言，即在编译和运行时进行大量的类型检查，以防止不匹配的数据类型发生。Java 也具有自动收集垃圾的功能，能够防止内存分配错误。同时，Java 还具有异常处理机制，使用 try/catch/finally 语句，程序员可以很快找到错误的代码，简化了错误处理与任务的恢复。

安全性：Java 没有 C++ 的指针操作，程序运行时，内存由系统分配，杜绝了非法的内存访问，这样可以避免病毒入侵。自动收集垃圾功能能够防止内存丢失、动态内存分配导致的问题。

可移植性：Java 先编译成字节码文件，然后由虚拟机解释执行，而不同操作系统都有各自对应的 JVM 版本。任何一台计算机只要安装了解释器（JVM 的一部分，用来解释 Java 编译后的程序），就可以运行 Java 程序，并且产生同样的结果。

高性能：Java 源程序在编译器中进行编译，将其转化为字节码，并在解释器中执行。Java 提供了一种"准实时"（Just In Time，JIT）编译器，在需要更快的速度时，可以使用 JIT 编译器将字节码转化成机器码，然后将其缓冲。

多线程：线程是一种轻量级进程，是比传统进程更小的可并发执行的单位。多线程处理能力使程序具有更好的交互性、实时性。Java 提供了多种实现多线程的方式，例如，继承 Thread 类、实现 Runnable 接口（无返回值，代码简洁）、实现 Callable 接口（有返回值，代码稍微复杂一点）。由于 Java 的单继承模式，所以通常情况下不会考虑第一种方式实现，针对后两者，如果无返回值，建议使用第二种方式（使用较多）；如果有返回值，可以选择第三种方式。

动态：在 Java 中，可以简单、直观地查询运行时的信息，也可以将新代码加入一个正在运行的程序。

16.2　Java 语法

16.2.1　数据类型

Java 基本数据类型可以简单概括为 4 类 8 种。

1. 整数类型

byte：8 位，有符号的整数，用于表示最小数据单位，取值范围为 -128~127，默认值为 0。

short：16 位，有符号的整数，取值范围为 -32768~32767，默认值为 0。

int：32 位，有符号的整数，是最常用的整数类型，取值范围为 -2^{31}~$2^{31}-1$，默认值为 0。

long：64 位，有符号的整数，取值范围为 -2^{63}~$2^{63}-1$，默认值为 0L。例：long a = 100000L。"L"理论上是不区分大小写的，但是为了与数字"1"区分开，最好还是大写。

2. 浮点类型

float：单精度类型，32 位，取值范围为 −3.403E38~3.403E38，默认值为 0.0f，"f"不区分大小写。

double：双精度类型，64 位，比较常用，取值范围为 −1.798E308~1.798E308，默认值为 0.0d。

3. 字符类型

char：char 类型是一个单一的 16 位 Unicode 字符。

4. 布尔类型

boolean：表示一位的信息，只有两个取值，true 和 false。

16.2.2　标识符

Java 的类名、变量名及方法名都称为标识符。关于标识符有以下几点需要注意：

（1）标识符必须由字母（A～Z,a～z）、美元符号（$）、下划线（_）开头；

（2）首字符之后可以由字母、数字、美元符号、下划线任意组合；

（3）标识符不可以是 Java 的关键字或保留字；

（4）标识符是区分大小写的。

建议：

（1）命名类名时每个单词首字母大写，如 HelloWorld；

（2）变量名和方法名通常首字母小写，采用驼峰命名法，如 helloWorld；

（3）常量命名通常是所有字母全大写，如 MAX。有多个单词时建议用下划线隔开，如 MAX_SIZE。

16.2.3　修饰符

Java 的修饰符主要用来修饰类、方法和属性，有两类修饰符。

1. 访问控制修饰符

包括 public、protected、default（没有修饰符）、private，如表 16-1 所示。

表 16-1

访问级别	访问控制修饰符	同类	同包	子类	不同包
公开	plbilc	√	√	√	√
受保护	protected	√	√	√	×
默认	defauct	√	√	×	×
私有	private	√	×	×	×

2. 非访问控制修饰符

包括 final、abstract、static、volatile、synchronized、native、transient。

16.2.4 变量

类变量：独立于方法之外的变量，用 static 修饰。

实例变量：独立于方法之外的变量，不用 static 修饰。

局部变量：方法中的变量。

如图 16-1 所示。

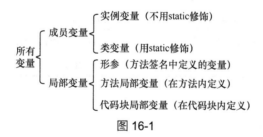

图 16-1

16.2.5 运算符

1. 算术运算符

算术运算符如表 16-2 所示。

表 16-2　算术运算符

运算符	运算	范例	结果
+	正号	+3	3
-	负号	b=4;-b	-4
+	加	5+5	10
-	减	6-4	2

续表

运算符	运算	范例	结果
*	乘	3*4	12
/	除	5/5	1
%	取模	7%5	2
++	自增（前）：先运算后取值	a=2;b=++a	a=3;b=3
++	自增（后）：先取值后运算	a=2;b=a++	a=3;b=2
--	自减（前）：先运算后取值	a=2;b=--a	a=1;b=1
--	自减（后）：先取值后运算	a=2;b=a--	a=1;b=2
+	字符串相加	"He"+"llo"	"Hello"

2. 赋值运算符

赋值运算符主要包括 =、+=、-=、/=、%=，如表 16-3 所示。

表 16-3　赋值运算符

运算符	运算	范例	结果
=	等于：直接赋值	c=1+2	c=3
+=	加等于：先运算再赋值	a=1;b=2;b+=a(等价于 b=b+a)	b=3
-=	减等于：先运算再赋值	a=1;b=2;b-=a(等价于 b=b-a)	b=1
/=	除等于：先运算再赋值	a=1;b=2;b/=a(等价于 b=b/a)	b=2
%=	取模等于：先运算再赋值	a=2;b=3;b%=a(等价于 b=b%a)	b=1

3. 比较运算符

比较运算符如表 16-4 所示。

表 16-4　比较运算符

运算符	运算	范例	结果
==	相等	4==3	false
!=	不等于	4!=3	true
<	小于	4<3	false
>	大于	4>3	true
<=	小于等于	4<=3	false
>=	大于等于	4>=3	true
instanceof	检查是否是类的对象	"Hello" instanceof String	true

4. 逻辑运算符

逻辑运算符如表 16-5 所示。

表 16-5 逻辑运算符

a	b	a&b （逻辑与）	a|b （逻辑或）	!a （取反）	a^b （异或）	a&&b （短路与）	a||b （短路或）
true	true	true	true	false	false	true	true
true	false	false	true	false	true	false	true
false	true	false	true	true	true	false	true
false	false	false	true	true	false	false	false

5. 位运算符

位运算符及其细节如表 16-6、表 16-7 所示。

表 16-6 位运算符

运算符	运算	范例
<<	左移	3<<2=12
>>	右移	3>>1=1
>>>	无符号右移	3>>>1=1
&	与运算	6&3=2
|	或运算	6|3=7
^	异或运算	6^3=5
~	反码	~6=-7

表 16-7 位运算符的细节

运算符	运算	说明
<<	左移	被移除的高位丢弃，空缺位补 0
>>	右移	被移位的二进制最高位是 0，右移后，空缺位补 0； 被移位的二进制最高位是 1，空缺位补 1
>>>	无符号右移	被移位的二进制最高位无论是 0 还是 1，空缺位都补 0
&	与运算	二进制位进行与运算，只有 1&1 时结果是 1，否则是 0
|	或运算	二进制位进行或运算，只有 0|0 时结果是 0，否则是 1
^	异或运算	相同二进制位进行异或运算，结果是 0，如 1^1=0,0^0=0 不相同二进制位异或运算，结果是 1，如 1^0=1,0^1=1
~	反码	正数取反，各二进制码按补码各位取反 负数取反，各二进制码按补码各位取反

6. 三元运算符

(条件表达式)? 表达式1：表达式2

条件表达式为 true，运算后结果为表达式1；

条件表达式为 false，运算后结果为表达式2。

16.2.6 关键字

本节主要讲解关键字，如表16-8所示。

表16-8 关键字

分类	关键字	说明
访问控制	public	公共的
	private	私有的
	protected	受保护的
类、修饰符	abstract	抽象
	class	类
	extends	继承
	implements	实现（接口）
	interface	接口
	enum	枚举
	final	最终的，不可改变的
	native	本地，原生方法
	new	创建
	static	静态
	strictfp	严格，精准
	transient	短暂的
	volatile	易变的
	synchronized	同步线程
程序控制	for	循环
	break	跳出循环
	continue	继续下一次循环
	do	运行
	while	循环
	switch	选择

续表

分类	关键字	说明
程序控制	case	定一个值供 switch 选择
	default	默认
	if	如果
	else	否则
	return	返回
异常处理	try	捕获异常
	catch	捕捉异常
	finally	必执行
	throw	抛出一个异常
	throws	声明一个异常可能被抛出
	assert	断言
包	import	引入
	package	包
数据类型	byte	字节型
	short	短整型
	int	整型
	long	长整型
	float	单精度浮点型
	double	双精度浮点型
	char	字符型
	boolean	布尔型
变量引用	super	父类，超类
	this	本类
	void	无返回值
保留关键字	null	空
	const	是关键字，但不能使用
	goto	是关键字，但不能使用

16.2.7 注释

单行注释: // 注释内容。效果如图 16-2 所示。

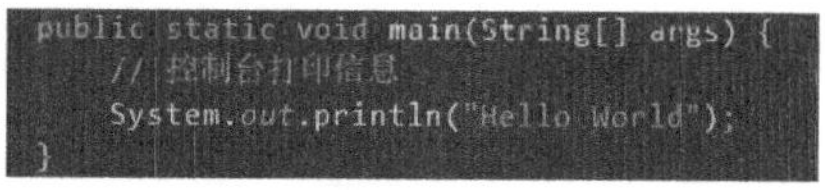

图 16-2

多行注释：/* 注释内容 */。效果如图 16-3 所示。

文档注释：/** 文档注释 */。效果如图 16-4 所示。

图 16-3　　　　　　　　　　　　图 16-4

16.3　Java 开发环境

16.3.1　JDK

JDK（Java Development Kit）是 Java 开发工具包，包含 JRE，同时也是 Java 开发运行的必备工具。如果安装了 JDK，就不用再安装 JRE 了。

16.3.2　JRE

JRE（Java Runtime Environment）是 Java 的运行环境，包含 JVM 及 Java 的核心类库。

16.3.3　JVM

JVM（Java Virtual Machine）是一种用于计算机设备的规范。它是虚拟出来的计算机，通过在实际的计算机上仿真模拟各种计算机功能来实现。

16.3.4　配置环境变量

（1）新建 JAVA_HOME 变量，变量值为 jdk 的安装目录。

（2）找到 Path 变量，进行编辑，在末尾添加 %JAVA_HOME%\bin;%JAVA_

HOME%\jre\bin;（注意：若原来 Path 变量的末尾没有分号，需要先在末尾添加分号。

（3）新建 CLASSPATH 变量，变量值为 .;%JAVA_HOME%\lib;%JAVA_HOME%\lib\tools.jar。

（4）检查是否配置成功。在命令指示符窗口输入 java -version，若出现图 16-5 所示的结果，说明配置成功。

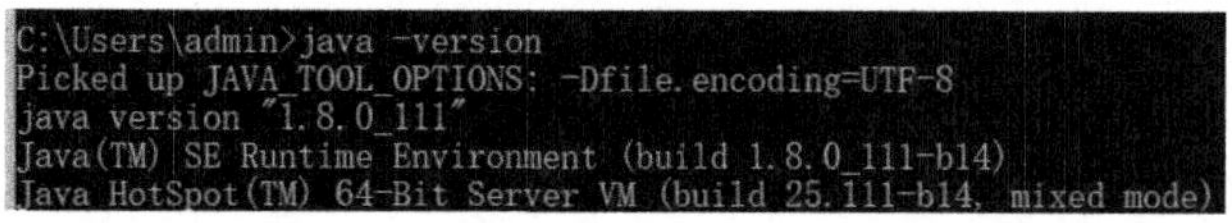

图 16-5

第 17 章　Spring Boot 框架

【本章导读】

◎ Spring Boot 概述

◎ Spring Boot 解决的问题

◎ Spring Boot 核心机制

◎ Spring Boot 优缺点

17.1　Spring Boot 概述

Spring 框架是 Java 平台的一种开源应用框架，它为开发者提供了一系列的解决方案，比如利用控制反转（IOC）的特性将对象统一管理；利用面向切面编程（AOP）的特性可以对业务的各个部门进行隔离，降低业务模块之间的耦合度。

Spring 核心特性如下。

1. IOC

IOC 全称为 Inversion of Control，中文译为控制反转，还有一个别名为 DI，全称为 Dependency Injection，中文译为依赖注入。IOC 通过 Java 反射机制对 Java 对象进行统一配置和管理，把对象的创建权由对象本身交给 Spring 容器，因此称为控制反转。同时 IOC 通过 Spring 注入对象，因此也称为依赖注入。

2. AOP

AOP 全称为 Aspect Oriented Programming，中文译为面向切面编程，是对面向对象编程（OOP）的补充。OOP 是将程序分解为对象，而 AOP 是将程序运行过程分解为各个切面。利用 AOP 可以对业务逻辑的各个部分进行隔离，从而降低业务逻辑各部分之间的耦合度，提高程序的可重用性，进而提高开发效率。

Spring Boot 是一个全新开源的轻量级框架。它基于 Spring4.0 设计，不仅继承了 Spring 框架原有的优秀特性，而且还简化了配置、项目搭建和开发。

Spring Boot 通过集成大量的框架解决了很多 Jar 包冲突及不稳定性等问题。

17.2 Spring Boot 解决的问题

1. 使编码变得简单

在 Spring 项目中，一般需要在 pom.xml 中添加多个依赖，而 Spring Boot 则会帮助开发者快速启动一个 Web 容器，只需要在 pom.xml 中添加如图 17-1 所示的依赖就可以。

```xml
<dependency>
    <groupId>org.springframework.boot</groupId>
    <artifactId>spring-boot-starter-web</artifactId>
</dependency>
```

图 17-1

点击此 pom.xml 进入该依赖后，发现里面已经包含之前在 Spring 项目中需要的依赖，如图 17-2 所示。

```xml
<dependency>
    <groupId>org.springframework.boot</groupId>
    <artifactId>spring-boot-starter</artifactId>
    <version>2.1.7.RELEASE</version>
    <scope>compile</scope>
</dependency>
<dependency>
    <groupId>org.springframework.boot</groupId>
    <artifactId>spring-boot-starter-json</artifactId>
    <version>2.1.7.RELEASE</version>
    <scope>compile</scope>
</dependency>
<dependency>
    <groupId>org.springframework.boot</groupId>
    <artifactId>spring-boot-starter-tomcat</artifactId>
    <version>2.1.7.RELEASE</version>
    <scope>compile</scope>
</dependency>
<dependency>
    <groupId>org.hibernate.validator</groupId>
    <artifactId>hibernate-validator</artifactId>
    <version>6.0.17.Final</version>
    <scope>compile</scope>
</dependency>
<dependency>
    <groupId>org.springframework</groupId>
    <artifactId>spring-web</artifactId>
    <version>5.1.9.RELEASE</version>
    <scope>compile</scope>
</dependency>
<dependency>
    <groupId>org.springframework</groupId>
    <artifactId>spring-webmvc</artifactId>
    <version>5.1.9.RELEASE</version>
    <scope>compile</scope>
</dependency>
```

图 17-2

2. 使配置变得简单

在 Spring 项目中，各种 XML、properties 配置文件，非常烦琐，也容易出错。但 Spring Boot 整个工程只有一个 application.properties（application.yml）文件，配置起来非常简单。

3. 使部署变得简单

在使用 Spring 项目并对项目部署时，我们需要在服务器或本地部署 tomcat 容器，然后把对应的 War 包放入 tomcat 中。但是 Spring Boot 不需要部署 tomcat，因为 Spring Boot 内嵌了 tomcat，我们只需要将项目打包成 Jar 包，使用 java -jar xxx.jar 启动项目。

4. 使监控变得简单

Spring Boot 工程可以直接引入 spring-boot-start-actuator 依赖，使用 REST 方式获取进程运行期间的性能参数，进而达到监控的目的，非常简便。

17.3 Spring Boot 核心机制

Spring Boot 主要有以下 4 大核心机制。

（1）auto-configuration：针对很多 Spring 应用程序常见的应用功能，Spring Boot 能自动提供相关配置。

（2）Starter 依赖：Spring Boot 需要什么功能，它就能引入相应需要的库。

（3）Spring Boot CLI：这个为可选特性，主要是针对 Groovy，简化了开发的流程。

（4）Spring Boot Actuator：能够深入运行中的 Spring Boot 应用程序，监控程序内部的各个信息。

17.4 Spring Boot 优缺点

1. 优点

（1）能够快速创建独立运行的 Spring 项目并与主流框架集成。

（2）使用嵌入式的 Servlet 容器，无须打包成 war 包，直接通过 main 方法启动。

（3）提供了 starter pom，能够非常方便地对 Jar 包进行管理，简化了 maven 配置。

（4）大量的自动配置，简化了开发。

（5）无须配置 XML，采用注解自动配置。

（6）提供了程序的监控。

（7）与云计算天然集成。

2. 缺点

（1）集成度太高，不知道底层实现。

（2）配置比较少，报错时很难定位。

第 18 章 后端工具 / 库

【本章导读】
◎ Java IDE
◎ Navicat
◎ Postman

18.1 Java IDE

IDE（Integrated Development Environment，集成开发环境）通常包括编辑器、构建工具、调试器等，部分 IDE 会有编译器 / 解释器。使用 IDE 可以让开发者提高开发效率，但并不建议初学者使用，初学者推荐使用 EditPlus 或者 Nodepad++ 编程。

18.1.1 常用的 IDE

1. Eclipse

Eclipse 是跨平台开源 IDE，由 IBM 公司创建，早期主要用于 Java 开发，但通过安装插件的方式也可以作为 C++、Python 的开发工具。Eclipse 支持各种插件安装，具有很高的灵活性。

2. MyEclipse

MyEclipse 是由 Genuitec 公司在 Eclipse 的基础上开发的一款 Java EE 集成开发环境，付费使用。不推荐使用这款 IDE。

3. NetBeans

NetBeans 是由 Sun 公司开发，开源，也支持多种语言开发，与 Eclipse 类似，但是使用率、流行程度不高。

4. IntelliJ IDEA

这款 IDE 是由 JetBrains 公司开发，于 2001 年推出，现在是市面上最优秀的 IDE

之一，不过大多都是收费软件，2009 年以后推出了免费的社区开源版本。推荐使用这款 IDE，网上也有各种版本的激活方法。

18.1.2 使用 IntelliJ IDEA 创建 Java 项目

（1）打开 IDEA，点击【Create New Project】，如图 18-1 所示。

图 18-1

（2）接着在弹出的对话框中选择【Java】，并选择 SDK，然后点击【Next】，如图 18-2 所示。

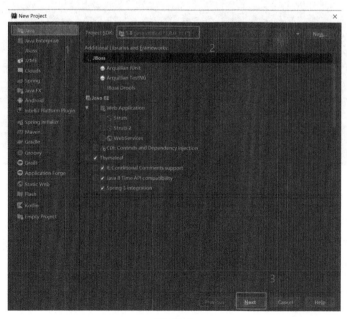

图 18-2

（3）接下来是选择生成 Java 文件，如果勾选【Java Hello World】，会生成一个默认的 Hello World 文件，也可以都不勾选，直接点击【Next】，如图 18-3 所示。

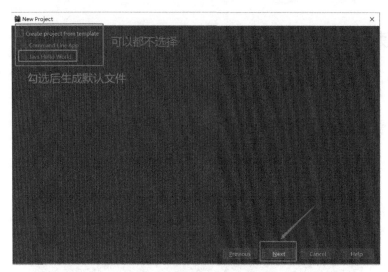

图 18-3

（4）下一步是给项目命名，然后点击【Finish】，如图 18-4 所示。

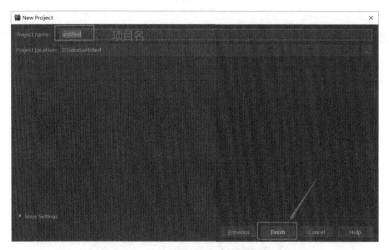

图 18-4

（5）创建完成，如图 18-5 所示。

（6）创建文件包，右击 src，选择【New】→【Package】，如图 18-6 所示。

（7）给文件包命名，如图 18-7 所示。

（8）新建 Class 文件，右击 demp，选择【New】→【Java Class】，如图 18-8 所示。

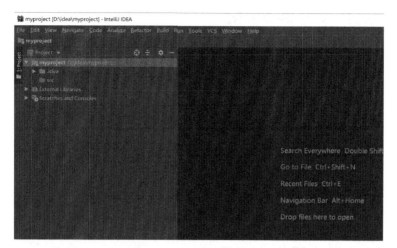

图 18-5

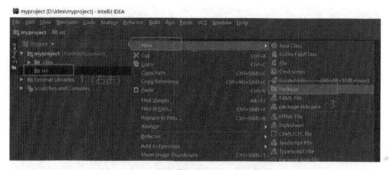

图 18-6

图 18-7

图 18-8

（9）给文件命名，如图 18-9 所示。

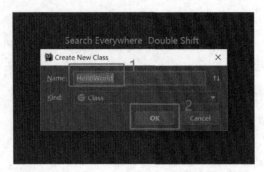

图 18-9

（10）编写主程序，如图 18-10 所示。

图 18-10

（11）开始运行，在编辑器内任意位置右击，选择 Run 'Hello World main()'，如图 18-11 所示。

（12）运行结果，如图 18-12 所示。

图 18-12

18.1.3 使用 IntelliJ IDEA 创建 Spring Boot 项目

（1）打开 IDEA，点击【Create New Project】，如图 18-13 所示。

图 18-13

（2）接着在弹出的对话框中选择【Spring Initializr】，并选择 SDK，默认勾选【Defanlt】，点击【Next】，如图 18-14 所示。

（3）给项目命名后，点击【Next】，如图 18-15 所示。

（4）选择插件，这里只需要勾选【Web】下的【Spring Web Start】就可以了，然后点击【Next】，如图 18-16 所示。

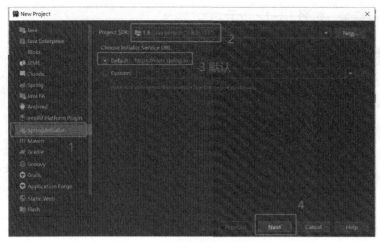

图 18-14

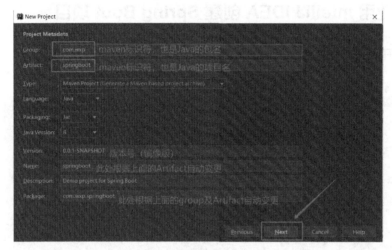

图 18-15

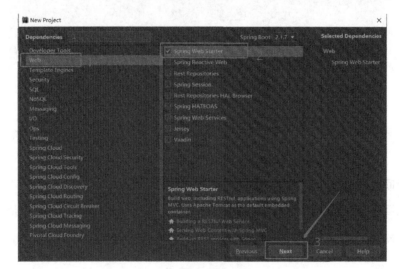

图 18-16

（5）确认项目名（一般不更改），直接点击【Finish】，如图 18-17 所示。

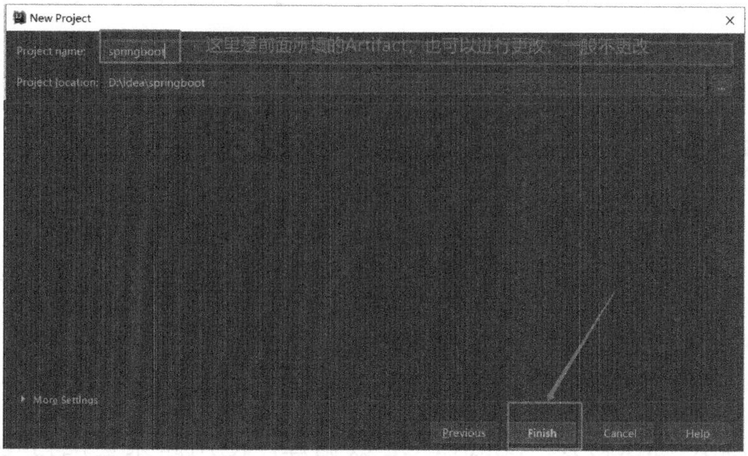

图 18-17

（6）创建完成，并删除无用文件，如图 18-18 所示。

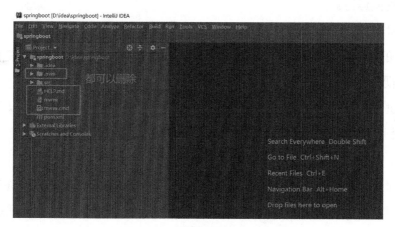

图 18-18

18.2　Navicat

Navicat 是一套快速、可靠的数据库管理工具，是为简化数据库管理而设计。它的设计符合数据库管理员（DBA）、开发人员及企业的需要，操作简便。

Navicat 提供 7 种语言供用户选择，被公认为全球最受欢迎的数据库前端用户界面工具之一。它可以用来对本机或远程的 MySQL、SQL Server、SQLite、Oracle 及

PostgreSQL 数据库进行管理及开发。Navicat 适用于 3 种操作系统，即 Windows、MacOS X 及 Linux Navicat。提供的功能有数据模型、数据传输、数据同步、结构同步、导入、导出、备份、还原、报表创建工具等。

18.2.1 版本

Navicat 主要提供了如下版本：Navicat Premium、Navicat for MySQL、Navicat for Oracle、Navicat for SQLite、Navicat for SQL Server、Navicat for PostgreSQL、Navicat Report Viewer、Navicat Data Modeler。

Navicat Premium 的功能非常强大，它可以通过单一程序同时连接 MySQL、SQL Server、SQLite、Oracle 及 PostgreSQL 数据库，使用起来更加方便。它也支持 MySQL、SQL Server、SQLite、Oracle 及 PostgreSQL 之间传输数据，这可以简化从一台服务器迁移数据到另一台服务器的类型的操作，不同数据库的批处理任务也可以计划并在给定时间运行。

Navicat for MySQL 是专门为 MySQL 设计的高性能数据库管理工具。它可以用于 3.21 或以上的 MySQL 版本，并支持大部分 MySQL 最新版本的功能，包括触发器、函数、事件、视图、管理用户等。

Navicat for Oracle 是专门为 Oracle 设计的数据库管理工具。它可以用于 8i 或以上的 Oracle 版本，并支持大部分 Oracle 最新版本的功能，包括目录、表空间、同义词、实体化视图、触发器、序列、类型等。

Navicat for SQLite 是专门为 SQLite 设计的数据库管理及开发工具。它可以用于 2 或 3 的 SQLite 版本，并支持大部分 SQLite 的功能，包括触发器、索引、视图等。

Navicat for SQL Server 是专门为 SQL Server 设计的高性能数据库开发及管理工具。它可以用于 SQL server 2000、2005、2008R2 及 SQL Azure，并支持大部分 SQL Server 的功能，包括触发器、索引、视图等。

Navicat for PostgreSQL 是专门为 PostgreSQL 设计的数据库管理及开发工具。它可以用于 7.5 或以上的 PostgreSQL 版本，并支持大部分 PostgreSQL 最新版本的功能，包括触发器、函数、管理用户等。

Navicat Report Viewer 是一个比较容易使用的工具，有友好的图形用户界面，可以在本机或远程操作各种数据库的报表。

Navicat Data Modeler 是一个强大且易于使用的数据库设计工具，用于创建和操作数据模型。它支持各种数据库，包括 MySQL、SQL Server、SQLite、Oracle 及 PostgreSQL。

18.2.2 Navicat for MySQL 使用

1. 连接 MySQL

打开 Navicat for MySQL，点击【连接】，输入对应的信息，如图 18-19 所示。

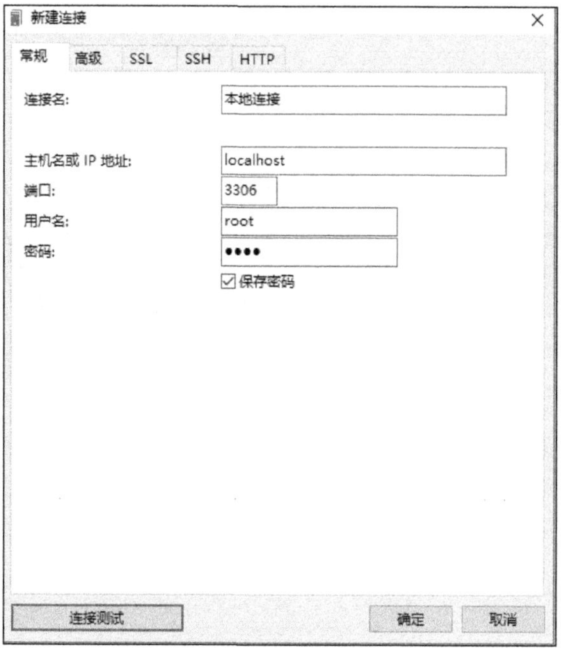

图 18-19

点击【连接测试】，如果成功，点击【确定】，如图 18-20 所示。

图 18-20

2. 新建数据库和表

鼠标右键单击【本地连接】，选择【新建数据库】，其中字符集一般选择 utf8，点击【确定】，如图 18-21 所示。

双击打开 test 数据库，右击新建表，编辑表字段，点击【保存】并输入表名，如图 18-22、图 18-23 所示。

3. 新建查询，编写 SQL 语句

插入行，主键 id 为自增类型，所以不需要考虑 id 列，输入语句 INSERT INTO test（'name'）VALUES（'test'），如图 18-24 所示。

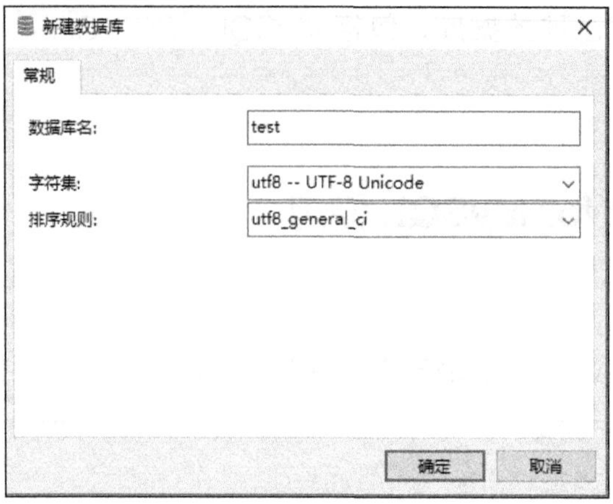

图 18-21

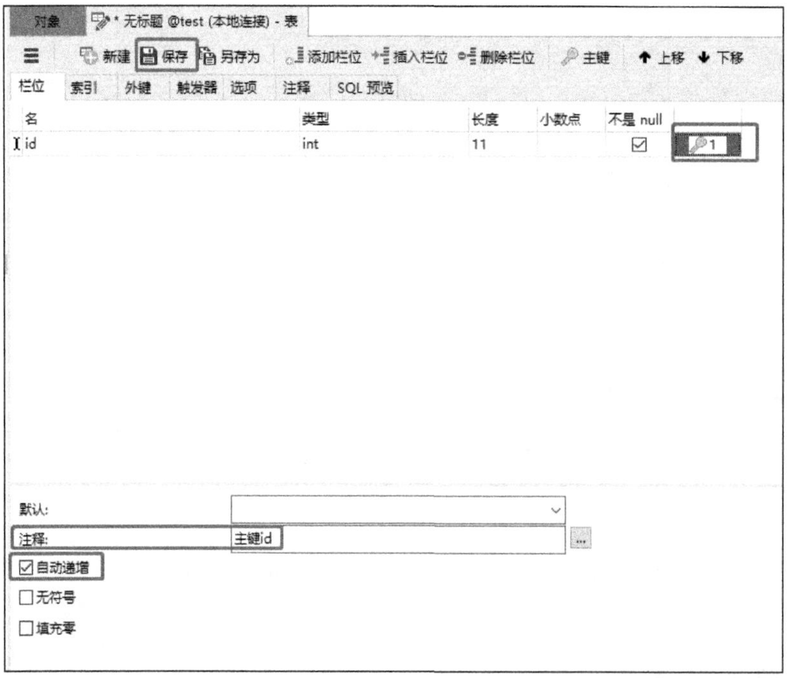

图 18-22

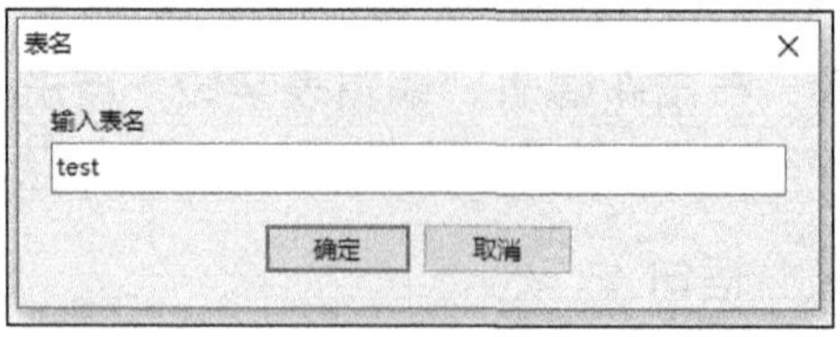

图 18-23

图 18-24

查询数据,输入语句 SELECT * FROM test;,如图 18-25 所示。Navicat 支持两种单行注释,"#"或者"--"都可以。/* */ 为多行注释。

图 18-25

修改行,输入语句 UPDATE test SET 'name' = 'test2' where id = 1;,如图 18-26 所示。

删除行,输入语句 DELETE FROM test where id = 1,如图 18-27 所示。

图 18-26

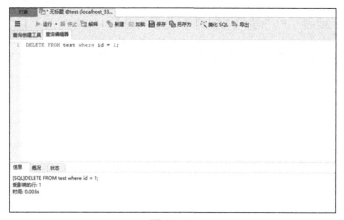

图 18-27

18.3 Postman

Postman 是一款功能强大的 Web API 与 HTTP 请求调试工具,Postman 能发送任何类型的 HTTP 请求,支持 GET/POST/PUT/DELETE 等,请求头中可以附带任何数量的 headers 信息。

Postman 使用非常简单，通过填写 URL、header、body 就可以发起一个请求，可以帮助我们做一些简单的接口调试及测试。

18.3.1 安装

Postman 有 2 种安装方式。

（1）通过 Chrome 浏览器插件的方式安装。在 Google 应用商店中搜索 Postman，然后添加至扩展程序。

（2）在 Postman 的官方网站下载安装程序，支持 Windows、macOS、Linux 等。下载完成后直接安装即可。

18.3.2 版本

Postman 一共有以下 4 个版本。

（1）Chrome 插件版本。

（2）Windows 版本（32/64 位）。

（3）macOS 版本。

（4）Linux 版本（32/64 位）。

18.3.3 使用

Postman 界面如图 18-28 所示。

图 18-28

发起一个请求，POST 与 GET 基本差不多，POST 入参放在 Body 中，GET 入参放在 Params 中或 URL 上拼接。这里以 POST 请求为例，如图 18-29 所示。

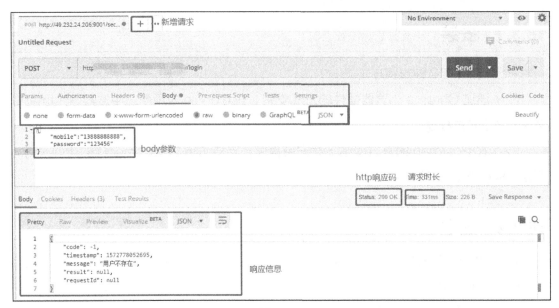

图 18-29

第 19 章　后端系统设计

【本章导读】

◎ 数据库设计

◎ 系统功能模块设计

◎ 接口设计

19.1　数据库设计

数据库包含如下 12 张表。

（1）user 用户表，用于存放用户的注册及基础信息，包括用户编号、手机号、密码、昵称、签名、来源、用户头像、登录次数、最后登录时间、是否登录、状态、注册时间等信息。

（2）product 商品表，用于存放用户上传的商品信息，包括商品编号、商品图片、商品描述、商品价格、商品类别、商品发布人、商品发布时地理位置、发布时间等信息。

（3）product_type 商品类别表，用于存放商品的基础类别信息，包括商品类别编号、商品类别名称、状态、创建时间、创建人等信息。

（4）order 订单表，用于存放用户购买商品的订单信息，包括订单编号、购买人编号、商品编号、交易状态、支付状态、购买状态、卖出状态、订单金额、支付金额、支付时间、订单完成时间、第三方支付流水号、订单创建时间、收货人手机号码、收货人姓名、收货人地址等信息。

（5）payment_log 支付流水表，用于存放支付流水信息，包括订单编号、支付请求订单号、第三方支付流水号、订单交易状态、交易金额、实际支付金额、用户编号、支付渠道、支付请求时间、实际完成时间等信息。

（6）payment_notify_log 支付回调流水表，用于存放支付回调信息，包括第三方支付订单号、系统订单号、支付状态、回调请求入参、请求时间等信息。

（7）praise 点赞表，用于存放用户对商品的点赞信息，包括商品编号、点赞用户编号、昵称、头像、点赞状态、点赞时间 / 去掉点赞时间等信息。

（8）comment_reply 评论回复表，用于存放用户的评论、回复信息，包括商品编号、评论 / 回复内容、评论 / 回复用户编号、评论 / 回复用户昵称、评论 / 回复用户头像、评论 / 回复时间、类型、回复人编号、回复人昵称、回复人头像、回复编号等信息。

（9）chat 聊天主表，用于存放聊天时用户、商品的关联信息，包括用户编号、对方用户编号、商品编号等信息。

（10）chat_list 聊天列表，用于存放聊天对话框信息，包括聊天主表编号、商品编号、当前用户编号、当前用户昵称、当前用户头像、对方用户编号、对方用户昵称、对方用户头像、是否在线、未读数、状态等信息。

（11）chat_detail 聊天详情表，用于存放聊天记录信息，包括聊天对话框编号、消息发送者用户编号、消息发送者用户昵称、消息发送者用户头像、发送内容、消息类型等信息。

（12）request_log 请求流水表，用于存放系统所有请求流水信息，包括请求方 ip、请求路径、头部信息、请求类型、请求参数、请求时间、异常信息、日志级别、请求时长等信息。

具体建表语句如下。

（1）用户信息表

```
CREATE TABLE 'user' (
  'user_id' varchar(32) NOT NULL COMMENT '用户编号',
  'mobile' varchar(20) DEFAULT NULL COMMENT '手机号',
  'password' varchar(100) DEFAULT NULL COMMENT '密码',
  'user_name' varchar(50) DEFAULT NULL COMMENT '昵称',
  'sign' varchar(255) DEFAULT NULL COMMENT '签名',
  'source' tinyint(4) DEFAULT NULL COMMENT '来源 1——H5',
  'user_avatar' varchar(150) DEFAULT NULL COMMENT '用户头像',
  'login_counts' int(6) DEFAULT '0' COMMENT '登录次数',
  'last_login_time' datetime DEFAULT NULL COMMENT '最后登录时间',
  'login_status' tinyint(2) DEFAULT '0' COMMENT '是否登录 1——已登录 2——未登录',
  'status' tinyint(2) DEFAULT '1' COMMENT '状态 1——正常 2——锁定',
  'create_time' datetime DEFAULT NULL COMMENT '创建时间（注册时间）',
  'update_time' datetime DEFAULT NULL COMMENT '修改时间',
  'remark' varchar(255) DEFAULT NULL COMMENT '备注',
```

```
    'device_id' varchar(500) DEFAULT NULL COMMENT '设备号',
    'client_version' varchar(100) DEFAULT NULL COMMENT '客户端版本号',
    'token' varchar(100) DEFAULT NULL COMMENT '用户令牌',
    'token_expired' bigint(20) DEFAULT NULL COMMENT 'token失效时间',
    'address' varchar(500) DEFAULT NULL COMMENT '收货地址',
    PRIMARY KEY ('user_id')
) ENGINE=InnoDB DEFAULT CHARSET=utf8 COMMENT='用户信息表';
```

（2）商品表

```
CREATE TABLE 'product' (
    'id' int(11) NOT NULL AUTO_INCREMENT COMMENT '商品编号',
    'product_imgs' varchar(2000) DEFAULT NULL COMMENT '商品图片,多个图片地址用英文逗号隔开',
    'product_desc' varchar(500) DEFAULT NULL COMMENT '商品描述',
    'product_price' decimal(10,2) DEFAULT NULL COMMENT '商品价格',
    'product_type_id' int(6) DEFAULT NULL COMMENT '商品类别',
    'publish_user_id' varchar(32) DEFAULT NULL COMMENT '发布人',
    'product_address' varchar(500) DEFAULT NULL COMMENT '商品发布时的地理位置',
    'create_time' datetime DEFAULT NULL COMMENT '创建时间（发布时间）',
    'update_time' datetime DEFAULT NULL COMMENT '修改时间',
    'remark' varchar(255) DEFAULT NULL COMMENT '备注',
    PRIMARY KEY ('id')
) ENGINE=InnoDB AUTO_INCREMENT=1 DEFAULT CHARSET=utf8 COMMENT='商品表';
```

（3）商品类别表

```
CREATE TABLE 'product_type' (
    'id' int(6) NOT NULL AUTO_INCREMENT COMMENT '商品类别编号',
    'name' varchar(100) DEFAULT NULL COMMENT '商品类别名称',
    'status' tinyint(2) DEFAULT NULL COMMENT '状态 1——有效; 2——无效',
    'create_time' datetime DEFAULT NULL COMMENT '创建时间',
    'create_man' varchar(50) DEFAULT NULL COMMENT '创建人',
    'update_time' datetime DEFAULT NULL COMMENT '修改时间',
    'update_man' varchar(50) DEFAULT NULL COMMENT '修改人',
    'remark' varchar(255) DEFAULT NULL COMMENT '备注',
    PRIMARY KEY ('id')
) ENGINE=InnoDB AUTO_INCREMENT=1 DEFAULT CHARSET=utf8 COMMENT='商品类别表';
```

（4）订单表

```sql
CREATE TABLE 'order' (
    'order_id' int(11) NOT NULL AUTO_INCREMENT COMMENT '订单编号',
    'buying_user_id' varchar(32) DEFAULT NULL COMMENT '购买人id',
    'product_id' int(11) DEFAULT NULL,
    'trade_status' tinyint(2) DEFAULT '1' COMMENT '交易状态 1——已下单 2——已取消 3——已结算',
    'pay_status' tinyint(2) DEFAULT '1' COMMENT '支付状态 1——待付款 2——付款中 3——已付款 4——付款失败',
    'buying_status' tinyint(2) DEFAULT '1' COMMENT '购买状态 1——有效 2——删除',
    'selling_status' tinyint(2) DEFAULT '1' COMMENT '卖出状态 1——有效 2——删除',
    'order_amount' decimal(15,2) DEFAULT NULL COMMENT '订单金额',
    'pay_amount' decimal(15,2) DEFAULT NULL COMMENT '支付金额',
    'pay_time' datetime DEFAULT NULL COMMENT '支付时间',
    'completion_time' datetime DEFAULT NULL COMMENT '订单完成时间',
    'out_trade_no' varchar(100) DEFAULT NULL COMMENT '第三方支付流水号',
    'create_time' datetime DEFAULT NULL COMMENT '订单创建时间',
    'update_time' datetime DEFAULT NULL COMMENT '修改时间',
    'remark' varchar(255) DEFAULT NULL COMMENT '订单备注',
    PRIMARY KEY ('order_id')
) ENGINE=InnoDB AUTO_INCREMENT=1 DEFAULT CHARSET=utf8 COMMENT='订单表';
```

（5）支付流水表

```sql
CREATE TABLE 'payment_log' (
    'id' int(11) NOT NULL AUTO_INCREMENT COMMENT '自增主键',
    'order_id' int(11) DEFAULT NULL COMMENT '商品订单号',
    'trade_no' varchar(100) DEFAULT NULL COMMENT '支付请求订单号',
    'out_trade_no' varchar(100) DEFAULT NULL,
    'trade_status' varchar(50) DEFAULT NULL COMMENT '订单交易状态',
    'trade_amount' decimal(15,2) DEFAULT NULL COMMENT '交易金额',
    'actual_pay_amount' decimal(15,2) DEFAULT NULL COMMENT '实际支付金额',
    'user_id' varchar(32) DEFAULT NULL COMMENT '用户编号',
    'pay_channel' tinyint(4) DEFAULT NULL COMMENT '支付渠道 1——支付宝 2——微信',
    'request_time' datetime DEFAULT NULL COMMENT '支付请求时间',
    'completion_time' datetime DEFAULT NULL COMMENT '实际完成时间',
```

```sql
  'request_params' varchar(2000) DEFAULT NULL,
  PRIMARY KEY ('id')
) ENGINE=InnoDB AUTO_INCREMENT=1 DEFAULT CHARSET=utf8 COMMENT='支付流水表';
```

（6）支付回调通知流水表

```sql
CREATE TABLE 'payment_notify_log' (
  'id' int(11) NOT NULL AUTO_INCREMENT COMMENT '自增主键',
  'trade_no' varchar(100) DEFAULT NULL COMMENT '第三方支付订单号',
  'out_trade_no' varchar(100) DEFAULT NULL COMMENT '系统订单号',
  'trade_status' varchar(50) DEFAULT NULL COMMENT '支付状态',
  'request_params' varchar(2000) DEFAULT NULL COMMENT '回调请求入参',
  PRIMARY KEY ('id')
) ENGINE=InnoDB AUTO_INCREMENT=1 DEFAULT CHARSET=utf8 COMMENT='支付回调通知流水表';
```

（7）点赞表

```sql
CREATE TABLE 'praise' (
  'id' int(11) NOT NULL AUTO_INCREMENT,
  'product_id' int(11) DEFAULT NULL COMMENT '商品编号',
  'user_id' varchar(32) DEFAULT NULL COMMENT '点赞用户编号',
  'user_name' varchar(50) DEFAULT NULL COMMENT '昵称',
  'user_avatar' varchar(255) DEFAULT NULL COMMENT '头像',
  'status' tinyint(2) DEFAULT '1' COMMENT '点赞状态 1——点赞 2——取消赞',
  'praise_time' datetime DEFAULT NULL COMMENT '点赞时间/取消点赞时间',
  PRIMARY KEY ('id')
) ENGINE=InnoDB AUTO_INCREMENT=1 DEFAULT CHARSET=utf8 COMMENT='点赞表';
```

（8）评论回复表

```sql
CREATE TABLE 'comment_reply' (
  'id' int(11) NOT NULL AUTO_INCREMENT COMMENT '评论编号主键',
  'product_id' int(11) DEFAULT NULL COMMENT '商品编号',
  'content' varchar(500) DEFAULT NULL COMMENT '评论内容',
  'from_user_id' varchar(32) DEFAULT NULL COMMENT '评论/回复用户编号',
```

```
    'from_user_name' varchar(50) DEFAULT NULL COMMENT '评论/回复用户昵称',
    'from_user_avatar' varchar(255) DEFAULT NULL COMMENT '评论/回复用户头像',
    'create_time' datetime DEFAULT NULL COMMENT '评论/回复时间',
    'type' tinyint(2) DEFAULT NULL COMMENT '类型 1-评论 2-回复',
    'to_user_id' varchar(32) DEFAULT NULL COMMENT '回复人编号',
    'to_user_name' varchar(50) DEFAULT NULL COMMENT '回复人昵称',
    'to_user_avatar' varchar(255) DEFAULT NULL COMMENT '回复人头像',
    'reply_id' int(11) DEFAULT NULL COMMENT '回复编号',
    PRIMARY KEY ('id')
) ENGINE=InnoDB AUTO_INCREMENT=1 DEFAULT CHARSET=utf8 COMMENT='评论表';
```

（9）聊天主表

```
CREATE TABLE 'chat' (
    'id' varchar(32) NOT NULL COMMENT '聊天主表编号',
    'user_id' varchar(32) DEFAULT NULL COMMENT '用户编号',
    'another_user_id' varchar(32) DEFAULT NULL COMMENT '对方用户编号',
    'product_id' int(11) DEFAULT NULL COMMENT '商品编号',
    PRIMARY KEY ('id')
) ENGINE=InnoDB DEFAULT CHARSET=utf8 COMMENT='聊天主表';
```

（10）聊天列表

```
CREATE TABLE 'chat_list' (
    'id' int(11) NOT NULL AUTO_INCREMENT COMMENT '主表自增主键',
    'chat_id' varchar(32) DEFAULT NULL COMMENT '聊天主表主键',
    'product_id' int(11) DEFAULT NULL COMMENT '商品编号',
    'user_id' varchar(32) DEFAULT NULL COMMENT '当前用户编号',
    'user_name' varchar(50) DEFAULT NULL COMMENT '当前用户昵称',
    'user_avatar' varchar(255) DEFAULT NULL COMMENT '当前用户头像',
    'another_user_id' varchar(32) DEFAULT NULL COMMENT '对方用户编号',
    'another_user_name' varchar(50) DEFAULT NULL COMMENT '对方用户昵称',
    'another_user_avatar' varchar(255) DEFAULT NULL COMMENT '对方用户头像',
    'is_online' tinyint(2) DEFAULT '2' COMMENT '是否在线 1——是 2——否',
    'unread' tinyint(4) DEFAULT '0' COMMENT '未读数',
    'status' tinyint(2) DEFAULT '1' COMMENT '状态 1——有效 2——删除',
    'create_time' datetime DEFAULT NULL COMMENT '创建时间',
    'update_time' datetime DEFAULT NULL COMMENT '更新时间',
```

```
    PRIMARY KEY ('id')
) ENGINE=InnoDB AUTO_INCREMENT=1 DEFAULT CHARSET=utf8 COMMENT='聊天
列表';
```

(11) 聊天详情表

```
CREATE TABLE 'chat_detail' (
    'id' int(11) NOT NULL AUTO_INCREMENT COMMENT '自增主键',
    'chat_id' varchar(32) DEFAULT NULL COMMENT '聊天主表编号',
    'user_id' varchar(32) DEFAULT NULL COMMENT '消息发送者用户编号',
    'user_name' varchar(50) DEFAULT NULL COMMENT '消息发送者用户昵称',
    'user_avatar' varchar(255) DEFAULT NULL COMMENT '消息发送者用户头像',
    'content' varchar(1000) DEFAULT NULL COMMENT '发送内容',
    'type' tinyint(2) DEFAULT '1' COMMENT '消息类型 1——用户消息 2——系
统消息',
    'is_latest' tinyint(2) DEFAULT '1' COMMENT '是否是最后一条消息 1——
是 2——否',
    'create_time' datetime DEFAULT NULL COMMENT '创建时间（发送时间）',
    'update_time' datetime DEFAULT NULL COMMENT '修改时间',
    PRIMARY KEY ('id')
) ENGINE=InnoDB AUTO_INCREMENT=1 DEFAULT CHARSET=utf8 COMMENT='聊天
详情表';
```

(12) 请求流水表

```
CREATE TABLE 'request_log' (
    'id' int(11) NOT NULL AUTO_INCREMENT COMMENT '自增主键',
    'ip' varchar(50) NOT NULL COMMENT '请求方ip',
    'url' varchar(500) DEFAULT NULL COMMENT '请求路径',
    'headers' varchar(2000) DEFAULT NULL COMMENT '头部信息',
    'request_type' varchar(50) DEFAULT NULL COMMENT '请求类型',
    'request_params' varchar(2000) DEFAULT NULL COMMENT '请求参数',
    'request_time' datetime DEFAULT NULL COMMENT '请求时间',
    'exception_detail' varchar(2000) DEFAULT NULL COMMENT '异常信息',
    'log_type' varchar(50) DEFAULT NULL COMMENT '日志级别 INFO;ERROR',
    'time' bigint(20) DEFAULT NULL COMMENT '请求时长',
    PRIMARY KEY ('id')
) ENGINE=InnoDB AUTO_INCREMENT=3869 DEFAULT CHARSET=utf8 COMMENT='
请求流水表';
```

19.2 系统功能模块设计

本项目模块分为用户模块、首页模块、商品模块、消息模块、个人中心模块以及支付模块。

用户模块：该模块主要是跟用户信息相关，用于注册、获取验证码、登录、校验是否登录。

首页模块：主要用于将最新的商品展示在首页进行轮播。

商品模块：该模块主要功能有获取商品类别列表、获取商品列表、发布商品、修改商品信息、获取商品详情、评论/回复、获取评论/回复列表、点赞/取消点赞、获取点赞列表等。

消息模块：该模块主要用于买家与卖家在线实时聊天，主要功能有发送消息、获取对话框详情、获取聊天对话框列表、初始化聊天等。

个人中心：该模块功能主要体现在客户端"我的"模块，功能包括获取商品列表、删除商品信息、更换手机号、找回密码、用户信息修改、商品数量查询、新增收货地址、修改收货地址、删除收货地址、获取收货地址列表、退出登录等。

支付模块：该模块功能主要包括购买商品、获取订单详情、取消订单、支付宝支付以及支付宝回调等。

具体模块及功能划分如表 19-1 所示。

表 19-1 系统接口清单

序号	模块	接口名称	接口功能
1	用户	register	注册
2		getVerifyCode	获取验证码
3		login	登录
4		checkLoginValid	校验是否登录
5	首次	getBannerList	获取 banner 列表
6	商品	getProductTypeList	获取商品类别列表
7		getProductList	获取商品列表
8		publishProduct	发布商品
9		updateProduct	修改商品
10		productDetails	获取商品详情
11		commentOrReply	评论/回复
12		getCommentReplyList	获取评论/回复列表

续表

序号	模块	接口名称	接口功能
13	商品	praiseOrUnPraise	点赞/取消点赞
14		getPraiseList	获取点赞列表
15	消息	getChatList	获取聊天对话框列表
16		getChatDetailList	获取聊天详情
17		initChat	初始化聊天
18		/socket/{userId}	发送消息
19	个人中心	getMyProductList	获取我的商品列表
20		delProduct	删除我的商品
21		verifyOldMobile	校验旧手机号
22		bindNewMobile	绑定新手机号
23		updatePwd	修改密码
24		retrievePwd	找回密码
25		updateUserInfo	用户信息修改
26		getProductNum	查询商品数量
27		logout	退出登录
28		getAddressList	获取收货地址列表
29		saveAddress	新增收货地址
30		updateAddress	修改收货地址
31		delAddress	删除收货地址
32	支付	placeOrder	购买商品
33		getOrderDetail	获取订单详情
34		cancelOrder	取消订单
35		alipay	支付宝支付
36		alipayNotify	支付宝支付回调

19.3 接口设计

本项目采用传统的 3 层开发架构，分别为 controller 层、service 层、dao 层。

其中 controller 层为控制层，主要作用是当用户接受客户端请求时，进行一些基础的参数校验，如是否为空校验、手机号格式校验、两次密码是否一致校验等。控制层的另一个作用是调用 service 层。

service 层为业务逻辑层,主要作用是处理业务,并调用 dao 层拿到数据,返回给 controller 层,最后返回给客户端。

dao 层为数据访问层,主要作用是与数据库交互,进行一些增删改查操作。

此外,本项目对接口的公共报文进行了统一处理,分别为公共请求报文、公共返回报文,具体报文信息如下。

(1)公共请求报文(统一放在 header 头部)如表 19-2 所示。

表 19-2 公共请求报文说明

参数名	必需	类型	字节长度	说明
token	否	string	32	用户令牌,用户登录后的唯一身份标识,登录后必传
requestId	否	string	32	请求流水号,会在公共返回报文带出
timestamp	否	long	13	请求时间戳
deviceId	否	string	500	设备号
clientVersion	否	string	100	客户端版本号

请求示例如下。

```
{
"token": "adfc9ba3211161e20190915221015",
"requestId": "d9b714de6306927201909022133016",
"timestamp": 1567301655000,
"deviceId": "iPhone 8 plus",
"clientVersion": "1.0.0"
}
```

(2)公共返回报文如表 19-3 所示。

表 19-3 公共返回报文说明

参数名	必需	类型	字节长度	说明
code	是	int	6	返回状态码
message	是	string	500	返回消息
result	是	string	2000	返回数据
requestId	否	string	32	请求报文的流水号
timestamp	是	long	13	当前时间戳

返回示例如下。

```
{
"code": 0,
"message": "success",
"result": null,
"requestId": "",
"timestamp": 1567301655000
}
```

code 返回值说明，如表 19-4 所示。

表 19-4　code 返回值说明

code 返回值	说明
0	成功
−1	失败
−2	参数异常
999	令牌失效

第 20 章 用户模块接口

【本章导读】

◎ 开发注册接口

◎ 开发获取验证码接口

◎ 开发登录接口

◎ 开发忘记密码接口

◎ 开发修改密码接口

◎ 开发用户信息修改接口

◎ 开发校验是否登录接口

◎ 开发退出登录接口

20.1 注册接口

1. 接口用途

此接口用于接收客户端上报的用户信息，用户信息包括：用户头像、用户昵称、手机号、密码等。通过程序校验并写入用户信息表。

2. 参数校验

校验参数是否为空；校验手机号格式、两次密码一致性。

3. 业务规则

通过查询 redis 存储的手机验证码，首先校验验证码是否有效，若无效则直接返回相应信息给客户端；若有效，继续通过查询数据库，判断手机号、昵称是否重复。如果重复则返回相应信息给客户端，保证手机号、昵称全表唯一。

4. 请求方式

采用 POST 请求。

5. 请求报文

注册接口请求报文如表 20-1 所示。

表 20-1　注册接口请求报文说明

参数名	必需	类型	字节长度	说明
userAvatar	是	file	—	用户头像
userName	是	string	50	用户昵称
mobile	是	string	20	用户手机号
verifyCode	是	string	6	短信验证码
password	是	string	100	密码
confirmPwd	是	string	100	确认密码
address	是	string	500	收货地址

6. 请求示例

POST 请求如图 20-1 所示。

图 20-1

7. 编写程序

（1）UserController 类，用户相关的接口控制层。此处代码主要校验参数是否为空、手机号格式、两次密码输入是否一致等，然后调用 service 层注册服务，并将报文返回给客户端。

```
/**
 *  注册接口
```

```
     * @param vo
     * 用户信息入参实体
     * @param userAvatar
     * 用户头像
     * @return
     */
    @RequestMapping(value = "register", method = RequestMethod.POST,
produces = MediaType.APPLICATION_JSON_UTF8_VALUE)
    public ResponseBO register(@RequestParam("userAvatar") MultipartFile
userAvatar, @Valid RegisterVO vo) {
        // 通过断言校验参数
        Assertion.notNull(userAvatar, "请选择图片");
        Assertion.isMobile(vo.getMobile(), "请输入正确的手机号");
        Assertion.isTrue(Detect.equals(vo.getPassword(), vo.getConfirmPwd
()), "两次密码输入不一致");
        // 调用业务逻辑层，注册服务
        userService.register(userAvatar, vo);
        return ResponseBO.success();
    }
```

（2）UserServiceImpl 类，用户相关的业务逻辑层。首先，通过手机号组装的 redis key 查询缓存中是否存在验证码，若不存在，说明验证码无效；若存在，继续对比客户端传入的验证码与缓存中的验证码是否相等，进而确认验证码是否正确。其次，通过查询数据库，判断用户昵称及手机号是否已经存在，接着上传用户头像。最后将用户信息组装好插入 user 表，完成注册业务。

```
    /**
     * 注册接口
     * @param file
     * @param vo
     */
    @Override
    public void register(MultipartFile file, RegisterVO vo) {
        String key = String.format(RedisConstants.REGISTER_VERIFY_CODE,
vo.getMobile());
        // 获取缓存中的验证码
        String code = RedisUtils.get(key);
        Assertion.notEmpty(code, "验证码已失效");
        Assertion.equals(vo.getVerifyCode(), code, "验证码错误");
```

```java
        // 判断用户是否已经注册
        Example example = new Example(User.class);
        Example.Criteria criteria = example.createCriteria();
        criteria.andEqualTo("mobile", vo.getMobile());
        List<User> list = userMapper.selectByExample(example);
        Assertion.isEmpty(list, "手机号已注册");
        example.clear();
        criteria = example.createCriteria();
        // 判断昵称是否已存在
        criteria.andEqualTo("userName", vo.getUserName());
        List<User> userNames = userMapper.selectByExample(example);
        Assertion.isEmpty(userNames, "昵称已存在");
        // 上传头像，OSS 存储
        OSSClient ossClient = OssUtil.getOssClient();
        String bucketName = OssUtil.getBucketName();
        String diskName = "images/user/" + DateUtils.getTimeString(new Date());
        // 根据当前时间戳及上传文件名生成最终的文件名
        String fileName = System.currentTimeMillis() + "_" + file.getOriginalFilename();
        boolean upload = false;
        try {
            upload = OssUtil.uploadInputStreamObject2Oss(ossClient, file.getInputStream(), fileName, bucketName, diskName);
        } catch (Exception e) {
            Assertion.isTrue(true, "头像上传失败");
        }
        // 上传成功
        if (upload) {
            // 用户头像地址
            String userAvatar = OssUtil.getOssUrl() + "/" + diskName + "/" + fileName;
            // 存储用户信息
            User user = new User();
            Date date = new Date();
            BeanUtils.copyProperties(vo, user);
            user.setUserId(UUIDUtils.getUid());
            user.setPassword(Md5Utils.md5(user.getPassword()));
            user.setCreateTime(date);
            user.setUpdateTime(date);
            user.setUserAvatar(userAvatar);
```

```
                userMapper.insertSelective(user);
                RedisUtils.del(key);
            } else {
                log.info("注册用户头像上传失败");
            }

        }
```

（3）OssUtil 类，阿里云 OSS 文件上传工具方法。

```
    /**
     * 向阿里云的 OSS 存储中存储文件，file 也可以用 InputStream 替代
     *
     * @param client
     * OSS 客户端
     * @param fileName
     * 上传文件
     * @param bucketName
     * bucket 名称
     * @param diskName
     * 上传文件的目录，bucket 下文件的路径
     * @return String 唯一 MD5 数字签名
     */
    public static boolean uploadInputStreamObject2Oss(OSSClient client,
InputStream is, String fileName, String bucketName, String diskName) {
        try {
            log.info("upload start");
            // 创建上传 Object 的 Metadata
            ObjectMetadata metadata = new ObjectMetadata();
            metadata.setContentLength(is.available());
            metadata.setCacheControl("no-cache");
            metadata.setHeader("Pragma", "no-cache");
            metadata.setContentEncoding("UTF-8");
            metadata.setContentType(getContentType(fileName));
            log.info("uploading" + bucketName + diskName + fileName);
            // 上传文件
            PutObjectResult ddd = client.putObject(bucketName, diskName
+ "/" + fileName, is, metadata);
            log.info("uploaded");
            log.info("上传文件 result :" + JSONObject.toJSONString(ddd));
```

```
        } catch (Exception e) {
            log.error(" 上传文件到 OSS 失败 ", e);
            return false;
        } finally {
            if (null != is) {
                log.info(" 关闭文件的输入流! ");
                try {
                    is.close();
                } catch (Exception e) {
                    log.error(" 关闭文件的输入流异常 ", e);
                }
                client.shutdown();
            }
        }
        return true;
    }
```

（4）UserMapper、dao 层、Spring Boot 与 MyBatis 无缝集成，导入 Jar 包后可直接继承，tk.mybatis.mapper.common.Mapper 类，单表增删改查无须写 SQL 及 interface 接口。

```
<dependency>
    <groupId>org.mybatis.spring.boot</groupId>
    <artifactId>mybatis-spring-boot-starter</artifactId>
    <version>2.1.0</version>
</dependency>
<dependency>
    <groupId>tk.mybatis</groupId>
    <artifactId>mapper-spring-boot-starter</artifactId>
    <version>2.1.5</version>
</dependency>

package com.xpwu.secondary.mapper;

import com.xpwu.secondary.entity.User;
import tk.mybatis.mapper.common.Mapper;

public interface UserMapper extends Mapper<User> {
}
```

（5）实体类 User，可以使用 MyBatis 生成工具，配置文件如下。

```java
package com.xpwu.secondary.entity;

import lombok.Getter;
import lombok.Setter;
import lombok.ToString;

import javax.persistence.Column;
import javax.persistence.Id;
import javax.persistence.Table;
import java.util.Date;

@Setter
@Getter
@ToString(callSuper = true)
@Table(name = "'user'")
public class User {
    /**
     * 用户编号
     */
    @Id
    @Column(name = "'user_id'")
    private String userId;

    /**
     * 手机号
     */
    @Column(name = "'mobile'")
    private String mobile;

    /**
     * 密码
     */
    @Column(name = "'password'")
    private String password;

    /**
     * 昵称
     */
```

```java
@Column(name = "'user_name'")
private String userName;

/**
 * 签名
 */
@Column(name = "'sign'")
private String sign;

/**
 * 来源 1——H5
 */
@Column(name = "'source'")
private Integer source;

/**
 * 用户头像
 */
@Column(name = "'user_avatar'")
private String userAvatar;

/**
 * 登录次数
 */
@Column(name = "'login_counts'")
private Integer loginCounts;

/**
 * 最后登录时间
 */
@Column(name = "'last_login_time'")
private Date lastLoginTime;

/**
 * 是否登录：1——已登录；2——未登录
 */
@Column(name = "'login_status'")
private Integer loginStatus;

/**
 * 状态：1——正常；2——锁定
```

```java
     */
    @Column(name = "'status'")
    private Integer status;

    /**
     * 创建时间（注册时间）
     */
    @Column(name = "'create_time'")
    private Date createTime;

    /**
     * 修改时间
     */
    @Column(name = "'update_time'")
    private Date updateTime;

    /**
     * 备注
     */
    @Column(name = "'remark'")
    private String remark;

    /**
     * 设备号
     */
    @Column(name = "'device_id'")
    private String deviceId;

    /**
     * 客户端版本号
     */
    @Column(name = "'client_version'")
    private String clientVersion;

    /**
     * 用户令牌
     */
    @Column(name = "'token'")
    private String token;

    /**
```

```
     * token 失效时间
     */
    @Column(name = "'token_expired'")
    private Long tokenExpired;

    /**
     * 收货地址
     */
    @Column(name = "'address'")
    private String address;

}
```

（6）入参实体 RegisterVO，这个实体主要是与请求报文一一对应。

```
package com.xpwu.secondary.vo;

import lombok.Getter;
import lombok.Setter;
import lombok.ToString;

import javax.validation.constraints.NotBlank;

/**
 * Created by IntelliJ IDEA.
 *
 * @author:caoxue
 * @date:2019/8/7 14:28
 * @description:注册接口入参
 * @version:1.0
 */
@Setter
@Getter
@ToString(callSuper = true)
public class RegisterVO extends BaseVO {

    private static final long serialVersionUID = 4370064067778524416L;

    /** 昵称 */
    @NotBlank(message = " 昵称不能为空 ")
```

```java
            private String userName;

            /** 手机号 */
            @NotBlank(message = "手机号不能为空")
            private String mobile;

            /** 验证码 */
            @NotBlank(message = "验证码不能为空")
            private String verifyCode;

            /** 密码 */
            @NotBlank(message = "密码不能为空")
            private String password;

            /** 确认密码 */
            @NotBlank(message = "确认密码不能为空")
            private String confirmPwd;

}
```

8. 返回报文及示例

参见公共返回示例。

20.2 获取验证码接口

1. 接口用途

此接口主要用于用户注册、找回密码、更换手机号时发送短信验证码。

2. 参数校验

校验参数是否为空；校验手机号格式、业务枚举类型、用户令牌（更换手机号）。

3. 业务规则

查询数据库，校验手机号是否已注册；验证码为 6 位随机数；有效时长 5min，验证码生成后通过手机号组装的 redis key 存入缓存中。

4. 请求方式

采用 POST 请求。

5. 请求报文

获取验证码接口请求报文如表 20-2 所示。

表 20-2 获取验证码接口请求报文说明

参数名	必需	类型	字节长度	说明
mobile	是	string	20	手机号
type	是	int	4	验证码类型：1——注册验证码，2——找回密码验证码，3——旧手机号验证，4——绑定新手机号

6. 请求示例

```
{
"mobile":"13888888888",
"type":1
}
```

7. 编写程序

（1）UserController 类，用户相关的接口控制层，用来接收客户端请求。getVerifyCode 方法主要用来校验参数是否为空、手机号格式以及验证码类型是否正确等，然后调用 service 层发送验证码服务，并将报文返回给客户端。

```
/**
 * 获取验证码接口
 * @param vo
 * @return
 */
@RequestMapping(value = "getVerifyCode", method = RequestMethod.POST,
produces = MediaType.APPLICATION_JSON_UTF8_VALUE)
public ResponseBO getVerifyCode(@Valid @RequestBody VerifyCodeVO vo) {
    Assertion.isMobile(vo.getMobile(), "请输入正确的手机号");
    Assertion.isTrue(VerifyCodeConstants.REGISTER == vo.getType()
            || VerifyCodeConstants.RETRIEVE_PWD == vo.getType()
            || VerifyCodeConstants.OLD_MOBILE == vo.getType()
            || VerifyCodeConstants.NEW_MOBILE == vo.getType(), "验证码类型错误");
    if (VerifyCodeConstants.OLD_MOBILE == vo.getType()) {
        // 校验 token 是否为空
```

```java
        String token = checkToken();
        vo.setToken(token);
    }
    userService.getVerifyCode(vo);
    return ResponseBO.success();
}
```

（2）VerifyCodeConstants 用于获取验证码业务类型常量，主要有注册获取验证码、找回密码获取验证码、验证旧手机号获取验证码、绑定新手机号获取验证码。

```java
public class VerifyCodeConstants {

    /**
     * 注册获取验证码
     */
    public final static int REGISTER = 1;

    /**
     * 找回密码获取验证码
     */
    public final static int RETRIEVE_PWD = 2;

    /**
     * 验证旧手机号获取验证码
     */
    public final static int OLD_MOBILE = 3;

    /**
     * 绑定新手机号获取验证码
     */
    public final static int NEW_MOBILE = 4;

}
```

（3）UserServiceImpl 类，getVerifyCode 方法用于发送验证码业务。当业务类型为注册获取验证码时，根据手机号查询是否注册，如果已注册，直接断言抛出异常信息"手机号已注册"，否则生成 6 位随机数，发送短信验证码，并将验证码存入 redis 中，有效时间 5min；当业务类型为找回密码获取验证码时，根据手机号查询是否已注册，若未注册，直接断言抛出异常信息，否则生成随机数，发送验证码；当业务类型为验证旧手机号获取验证码时，通过 token 拿到用户信息，若用户信息为空，直接断言抛出异常信息，否则继续判断旧手机号与用户信息里面的手机号是否一致，若不一致，抛出异常信息，否则发送验证码；当业务类型为绑定新手机号获取验证码时与注册一致。

```java
/**
 * 获取验证码
 * @param vo
 */
@Override
public void getVerifyCode(VerifyCodeVO vo) {
    String key;
    // 注册验证码
    if (VerifyCodeConstants.REGISTER == vo.getType()) {
        Example example = new Example(User.class);
        example.createCriteria().andEqualTo("mobile", vo.getMobile());
        User user = userMapper.selectOneByExample(example);
        Assertion.isNull(user, "手机号已注册");
        key = String.format(RedisConstants.REGISTER_VERIFY_CODE, vo.getMobile());
    } else if (VerifyCodeConstants.RETRIEVE_PWD == vo.getType()) {
        // 找回密码验证码
        Example example = new Example(User.class);
        example.createCriteria().andEqualTo("mobile", vo.getMobile());
        User user = userMapper.selectOneByExample(example);
        Assertion.notNull(user, "手机号未注册");
        key = String.format(RedisConstants.RETRIEVE_PWD_VERIFY_CODE, vo.getMobile());
    } else if (VerifyCodeConstants.OLD_MOBILE == vo.getType()) {
        // 校验旧手机号验证码
        User user = this.checkToken(vo.getToken());
        Assertion.notNull(user, "手机号未注册");
        Assertion.equals(user.getMobile(), vo.getMobile(), "手机号必须与注册手机号一致");
        key = String.format(RedisConstants.OLD_MOBILE_VERIFY_CODE, vo.getMobile());
    } else {
        Example example = new Example(User.class);
        example.createCriteria().andEqualTo("mobile", vo.getMobile());
        User user = userMapper.selectOneByExample(example);
```

```java
            Assertion.isNull(user, "手机号已注册");
            key = String.format(RedisConstants.NEW_MOBILE_VERIFY_CODE, vo.getMobile());
        }
        // 生成6位数验证码
        int code = (int) ((Math.random() * 9 + 1) * 100000);
        // 调用阿里云短信服务发送短信
        AliSmsUtils.getSms(vo.getMobile, code);
        // 将短信验证码放入 redis 并设置5分钟有效时间
        RedisUtils.setEx(key, RedisConstants.VERIFY_CODE_EXPIRE, String.valueOf(code));
        log.info("短信发送成功");
    }
```

（4）阿里云短信工具类 AliSmsUtils

```java
package com.xpwu.secondary.utils;

import com.alibaba.fastjson.JSON;
import com.aliyuncs.DefaultAcsClient;
import com.aliyuncs.IAcsClient;
import com.aliyuncs.exceptions.ClientException;
import com.aliyuncs.profile.DefaultProfile;
import com.aliyuncs.profile.IClientProfile;
import com.aliyuncs.sms.model.v20160927.SingleSendSmsRequest;
import com.aliyuncs.sms.model.v20160927.SingleSendSmsResponse;
import lombok.extern.slf4j.Slf4j;

/**
 * Created by IntelliJ IDEA.
 *
 * @author:caoxue
 * @date:2019/8/14 11:36
 * @description:阿里云短信发送工具类
 * @version:1.0
 */
@Slf4j
public class AliSmsUtils {

    private static String accessKeyId = PropertiesUtil.get("accessKeyId");
```

```java
    private static String secret = PropertiesUtil.get("secret");

    private static String endpointName = PropertiesUtil.get("endpointName");

    private static String regionId = PropertiesUtil.get("regionId");

    private static String product = PropertiesUtil.get("product");

    private static String domain = PropertiesUtil.get("domain");

    private static String signName = PropertiesUtil.get("signName");

    private static String templateCode = PropertiesUtil.get("templateCode");


    /**
     * 发送验证码
     *
     * @param phoneNum
     * @param code
     * @return
     */
    public static String getSms(String phoneNum, String code) {
        try {
            IClientProfile profile = DefaultProfile.getProfile(regionId,
                    accessKeyId, secret);
            DefaultProfile.addEndpoint(endpointName, regionId, product, domain);

            IAcsClient client = new DefaultAcsClient(profile);
            SingleSendSmsRequest request = new SingleSendSmsRequest();
            // 控制台创建的签名名称
            request.setSignName(signName);
            // 控制台创建的模板CODE
            request.setTemplateCode(templateCode);
            // 短信模板中的变量;数字需要转换为字符串;个人用户每个变量长度必须小于15个字符
            request.setParamString("{\"code\":\"" + code + "\"}");
            // 接收号码
            request.setRecNum(phoneNum);
            log.info("阿里云发送短信上送报文:{}", JSON.toJSONString(request));
```

```java
            SingleSendSmsResponse httpResponse = client.getAcsResponse(request);
            log.info("阿里云发送短信返回报文：{}", JSON.toJSONString(httpResponse));
            return httpResponse.getModel();
        } catch (ClientException e) {
            log.info("阿里云短信发送失败", e);
        }
        return null;
    }
```

（5）入参实体 VerifyCodeVO，入参为手机号及验证码类型。

```java
package com.xpwu.secondary.vo;

import lombok.Getter;
import lombok.Setter;
import lombok.ToString;

import javax.validation.constraints.NotBlank;
import javax.validation.constraints.NotNull;

/**
 * Created by IntelliJ IDEA.
 *
 * @author:caoxue
 * @date:2019/8/10 14:07
 * @description:发送短信验证码接口入参
 * @version:1.0
 */
@Setter
@Getter
@ToString(callSuper = true)
public class VerifyCodeVO extends BaseVO {

    private static final long serialVersionUID = -8940747596809818155L;

    /** 手机号 */
    @NotBlank(message = "手机号不能为空")
    private String mobile;
```

```
/** 验证码类型：1——注册验证码 2——找回密码验证码 */
@NotNull(message = "验证码类型不能为空")
private Integer type;
}
```

8. 返回报文及示例

参见公共返回示例。

20.3 登录接口

1. 接口用途

此接口用于已注册的用户通过手机号、密码登录。

2. 参数校验

校验手机号是否为空；校验密码是否为空。

3. 业务规则

根据手机号校验用户是否存在、验证密码是否正确、验证用户状态是否正常。登录成功后返回用户基础信息，包括用户令牌、昵称、头像、状态、上次登录时间、用户编号、手机号等。

4. 请求方式

采用 POST 请求。

5. 请求报文

登录接口请求报文如表 20-3 所示。

表 20-3　登录接口请求报文说明

参数名	必需	类型	字节长度	说明
mobile	是	string	20	手机号
password	是	string	20	登录密码

6. 请求示例

```
{
"mobile":"13888888888",
```

```
    "password":"123456"
}
```

7. 编写程序

（1）UserController 类，login 方法用于接收用户登录请求，通过 @Valid 注解、LoginVO 中 @NotBlank 注解校验手机号、密码是否为空，调用 service 层登录服务，并返回用户信息给客户端。

```
/**
 * 登录接口
 * @param vo
 * @return
 */
@RequestMapping(value = "login", method = RequestMethod.POST, produces = 
MediaType.APPLICATION_JSON_UTF8_VALUE)
public ResponseBO login(@Valid @RequestBody LoginVO vo) {
    return ResponseBO.success(userService.login(vo));
}
```

（2）UserServiceImpl 类，login 方法用于用户登录业务。首先通过手机号查询用户是否存在，若不存在，断言抛出异常信息，否则继续判断密码是否正确，若不正确，抛出异常信息，否则继续判断用户状态是否正常。若不正常，抛出异常信息，否则修改用户上次登录时间为当前时间，登录次数加 1，登录状态修改为已登录。然后根据用户编号生成 32 位 token，即用户令牌，并将 token 放入 redis 中，有效时间为 7 天，同时将用户信息也放入缓存中，有效时间为 7 天。最后将部分用户信息封装到 LoginBO 中返回给客户端。

```
/**
 * 登录
 * @param vo
 * @return
 */
@Override
public LoginBO login(LoginVO vo) {
    // 判断用户是否存在
    User old = checkUserExsits(vo.getMobile());
    Assertion.notNull(old, "用户不存在");
```

```java
        Assertion.equals(old.getPassword(), Md5Utils.md5(vo.getPassword
()), "密码不正确");
        Assertion.equals(1, old.getStatus(), "用户已被锁定");
        // 已存在用户信息,封装修改实体
        User entity = new User();
        Date date = new Date();
        entity.setUserId(old.getUserId());
        entity.setLastLoginTime(date);
        entity.setLoginCounts(old.getLoginCounts() + 1);
        entity.setLoginStatus(1);
        entity.setUpdateTime(date);
        // 生成token
        String token = TokenUtils.getToken(old.getUserId());
        // token失效时间点
        Long tokenExpired = System.currentTimeMillis() + RedisConstants.
TOKEN_EXPIRED * 1000;
        entity.setToken(token);
        entity.setTokenExpired(tokenExpired);
        userMapper.updateByPrimaryKeySelective(entity);
        // 封装token信息
        TokenBO tokenBO = new TokenBO();
        tokenBO.setUserId(old.getUserId());
        tokenBO.setMobile(old.getMobile());
        // 删除旧token
        RedisUtils.del(String.format(RedisConstants.TOKEN, old.getToken()));
        // 将token信息放入缓存并设置7天有效期
        RedisUtils.setEx(String.format(RedisConstants.TOKEN, token),
RedisConstants.TOKEN_EXPIRED, JSON.toJSONString(tokenBO));
        // 将最新用户信息放入缓存并设置7天有效期
        BeanUtil.copyPropertiesIgnoreNull(entity, old);
        RedisUtils.setEx(String.format(RedisConstants.USER_INFO, vo.
getMobile()), RedisConstants.USER_INFO_EXPIRE, JSON.toJSONString(old));
        // 封装返回客户端信息
        LoginBO bo = new LoginBO();
        BeanUtils.copyProperties(old, bo);
        bo.setLastLoginTime(date);
        bo.setToken(token);
        return bo;
    }
```

(3)根据用户编号生成token,主要逻辑为用户编号加上当前时间yyyyMMddHHmmss

组装后经过 MD5 加密算法，最后将所有字母大写，得到 32 位 token。

```java
    /**
     * 根据用户 id 及当前时间生成 token
     *
     * @param userId
     * @return
     */
    public static String getToken(String userId) {
        String md5Str = userId + DateUtils.getDate(new Date());
        return Objects.requireNonNull(MD5Utils.stringToMD5(md5Str)).toUpperCase();
    }
```

（4）入参实体 LoginVO，入参为手机号及登录密码。

```java
    package com.xpwu.secondary.vo;

    import lombok.Getter;
    import lombok.Setter;
    import lombok.ToString;

    import javax.validation.constraints.NotBlank;

    /**
     * Created by IntelliJ IDEA.
     *
     * @author:caoxue
     * @date:2019/8/10 10:41
     * @description:登录接口入参
     * @version:1.0
     */
    @Setter
    @Getter
    @ToString(callSuper = true)
    public class LoginVO extends BaseVO {

        private static final long serialVersionUID = -3644422290879382378L;
        /**
         * 手机号
```

```
     */
    @NotBlank(message = " 手机号不能为空 ")
    private String mobile;

    /**
     * 登录密码
     */
    @NotBlank(message = " 密码不能为空 ")
    private String password;

}
```

（5）出参实体 LoginBO 及返回报文实体。

```
package com.xpwu.secondary.bo;

import lombok.Getter;
import lombok.Setter;
import lombok.ToString;

import java.io.Serializable;
import java.util.Date;

/**
 * Created by IntelliJ IDEA.
 *
 * @author:caoxue
 * @date:2019/8/10 10:43
 * @description:登录成功返回参数
 * @version:1.0
 */
@Setter
@Getter
@ToString(callSuper = true)
public class LoginBO implements Serializable {

    private static final long serialVersionUID = 3653029165858373922L;

    /**
     * 用户令牌
```

```java
     */
    private String token;

    /**
     * 昵称
     */
    private String userName;

    /**
     * 用户头像
     */
    private String userAvatar;

    /**
     * 状态：1——正常；2——锁定
     */
    private Integer status;

    /**
     * 上次登录时间
     */
    private Date lastLoginTime;

    /**
     * 用户编号
     */
    private String userId;

    /**
     * 手机号
     */
    private String mobile;

    /**
     * 收货地址
     */
    private String address;

}
```

8. 返回报文

登录接口返回报文如表20-4所示。

表 20-4 登录接口返回报文说明

参数名	必需	类型	字节长度	说明
token	是	string	32	用户令牌
userName	是	string	20	用户昵称
userAvatar	是	string	100	用户头像
lastLoginTime	是	long	13	上次登录时间戳
userId	是	string	32	用户编号
status	是	int	2	用户状态：1——正常，2——锁定
mobile	是	string	20	手机号
address	是	string	500	收货地址

9. 返回示例

```
{
"code":0,
"message":"success",
"requestId":null,
"timestamp":1567301655000,
"result":{
  "token":"17303B76DC22B7638BDD4A47CA842A6F",
  "userName":" 测试 ",
  "userAvatar":"xxx",
  "status":1,
  "lastLoginTime":1567301655000,
  "userId":"25cd48b22898455ea49e6d5391e6774c",
  "mobile":"13888888888",
  "address":" 北京 "
  }
}
```

20.4 忘记密码接口

1. 接口用途

此接口用于用户忘记密码时，通过手机号重新设置新密码。

2. 参数校验

校验手机号、短信验证码、新密码、确认密码是否为空；校验新密码与确认密码是否一致。

3. 业务规则

根据手机号查询用户是否已经注册，从 redis 中取出验证码，验证是否正确，若正确，删除 redis 中的验证码信息。

4. 请求方式

采用 POST 请求。

5. 请求报文

忘记密码接口请求报文如表 20-5 所示。

表 20-5　忘记密码接口请求报文说明

参数名	必需	类型	字节长度	说明
mobile	是	string	20	手机号
verifyCode	是	string	6	短信验证码
newPwd	是	string	100	新密码
confirmPwd	是	string	100	确认密码

6. 请求示例

```
{
"mobile":"13888888888",
"verifyCode":"132940",
"newPwd":"123456",
"confirmPwd":"123456"
}
```

7. 编写程序

（1）UserController 类，forgotPwd 方法用于接收客户端忘记密码请求，进行参数校验，并调用 service 层服务，将相应报文信息返回给客户端。

```
/**
 * 忘记密码
```

```
 * @param vo
 * @return
 */
@RequestMapping(value = "forgotPwd", method = RequestMethod.POST,
produces = MediaType.APPLICATION_JSON_UTF8_VALUE)
public ResponseBO forgotPwd(@Valid @RequestBody ForgotPwdVO vo) {
    Assertion.equals(vo.getNewPwd(), vo.getConfirmPwd(), "两次密码输入不一致");
    userService.forgotPwd(vo);
    return ResponseBO.success();
}
```

（2）UserServiceImpl 类，forgotPwd 方法用于忘记密码业务。根据手机号查询用户是否已经注册，若未注册，直接返回异常信息，否则从 redis 中取出验证码，若取出的验证码字符串为空，说明验证码已失效。若取出的验证码与客户端传入的验证码不一致，则说明验证码错误，然后返回对应的错误信息，否则修改用户信息及同步 redis，并将缓存中的验证码删除。

```
/**
 * 忘记密码
 * @param vo
 */
@Override
public void forgotPwd(ForgotPwdVO vo) {
    Example example = new Example(User.class);
    example.createCriteria().andEqualTo("mobile", vo.getMobile());
    User user = userMapper.selectOneByExample(example);
    Assertion.notNull(user, "手机号未注册");
    String key = String.format(RedisConstants.RETRIEVE_PWD_VERIFY_CODE, vo.getMobile());
    // 获取缓存中的验证码
    String code = RedisUtils.get(key);
    Assertion.notEmpty(code, "验证码已失效");
    Assertion.equals(vo.getVerifyCode(), code, "验证码错误");
    User retrieve = new User();
    Date date = new Date();
    retrieve.setUserId(user.getUserId());
    retrieve.setPassword(Md5Utils.md5(vo.getNewPwd()));
```

```java
        retrieve.setUpdateTime(date);
        BeanUtil.copyPropertiesIgnoreNull(retrieve, user);
        this.updateUserInfo(retrieve, user);
        RedisUtils.del(key);
}
```

（3）入参实体 ForgotPwdVO，对应客户端请求报文。

```java
package com.xpwu.secondary.vo;

import lombok.Getter;
import lombok.Setter;
import lombok.ToString;

import javax.validation.constraints.NotBlank;

/**
 * Created by IntelliJ IDEA.
 *
 * @author:caoxue
 * @date:2019/8/10 23:55
 * @description:找回密码接口入参
 * @version:1.0
 */
@Setter
@Getter
@ToString(callSuper = true)
public class ForgotPwdVO extends BaseVO {
    private static final long serialVersionUID = 6130444602859371457L;

    /**
     * 手机号
     */
    @NotBlank(message = "手机号不能为空")
    private String mobile;

    /**
     * 短信验证码
     */
    @NotBlank(message = "短信验证码不能为空")
```

```
    private String verifyCode;

    /**
     * 新密码
     */
    @NotBlank(message = " 新密码不能为空 ")
    private String newPwd;

    /**
     * 确认密码
     */
    @NotBlank(message = " 确认密码不能为空 ")
    private String confirmPwd;

}
```

8. 返回报文及示例

参见公共返回示例。

20.5 修改密码接口

1. 接口用途

此接口用于用户修改密码。

2. 参数校验

检验用户令牌、旧密码、新密码、确认密码是否为空；检验新密码与确认密码是否一致。

3. 业务规则

根据 token 查询用户是否存在，即 token 是否有效；判断旧密码是否正确，校验新密码与旧密码是否一致。

4. 请求方式

采用 POST 请求。

5. 请求报文

修改密码接口请求报文如表 20-6 所示。

表 20-6 修改密码接口请求报文说明

参数名	必需	类型	字节长度	说明
oldPwd	是	string	100	旧密码
newPwd	是	string	100	新密码
confirmPwd	是	string	100	确认密码

6. 请求示例

```
{
"oldPwd":"123456",
"newPwd":"12345678",
"confirmPwd":"12345678"
}
```

7. 编写程序

（1）UserController 类，updatePwd 方法用于接收客户端修改密码请求。主要包含校验用户令牌，旧密码、新密码、确认密码是否为空，两次输入密码是否一致，然后调用 service 层修改密码业务，并返回报文信息给客户端。

```
/**
 * 修改密码
 * @param vo
 * @return
 */
@RequestMapping(value = "updatePwd", method = RequestMethod.POST,
produces = MediaType.APPLICATION_JSON_UTF8_VALUE)
public ResponseBO updatePwd(@Valid @RequestBody UpdatePwdVO vo) {
    // 校验token是否为空
    String token = checkToken();
    Assertion.equals(vo.getNewPwd(), vo.getConfirmPwd(), "两次密码输入不一致");
    vo.setToken(token);
    userService.updatePwd(vo);
    return ResponseBO.success();
}
```

（2）UserServiceImpl 类，updatePwd 方法用于修改密码业务。校验 token 有效

性并返回用户实体，若校验不通过，则返回相应的错误信息。否则继续判断旧密码是否正确，若不正确，则返回相应的错误信息。否则继续判断新密码与库中旧密码是否一样，若一样，则返回错误信息，否则修改数据库用户信息并同步 redis。

```java
/**
 * 修改密码
 * @param vo
 */
@Override
public void updatePwd(UpdatePwdVO vo) {
    // 校验token是否有效
    User user = this.checkToken(vo.getToken());
    Assertion.equals(user.getPassword(), Md5Utils.md5(vo.getOldPwd()), "旧密码不正确");
    Assertion.notEquals(user.getPassword(), Md5Utils.md5(vo.getNewPwd()), "新密码不能与旧密码一样");
    // 组装更新实体
    User update = new User();
    Date date = new Date();
    update.setUserId(user.getUserId());
    update.setUpdateTime(date);
    update.setPassword(Md5Utils.md5(vo.getNewPwd()));
    BeanUtil.copyPropertiesIgnoreNull(update, user);
    // 调用更新方法
    this.updateUserInfo(update, user);

}
```

（3）入参实体 UpdatePwdVO，对应客户端请求报文。

```java
package com.xpwu.secondary.vo;

import lombok.Getter;
import lombok.Setter;
import lombok.ToString;

import javax.validation.constraints.NotBlank;

/**
 * Created by IntelliJ IDEA.
```

```
 *
 * @author:caoxue
 * @date:2019/8/10 22:22
 * @description:修改密码接口入参
 * @version:1.0
 */
@Setter
@Getter
@ToString(callSuper = true)
public class UpdatePwdVO extends BaseVO {
    private static final long serialVersionUID = 18437081375559454805L;

    /**
     * 旧密码
     */
    @NotBlank(message = " 旧密码不能为空 ")
    private String oldPwd;

    /**
     * 新密码
     */
    @NotBlank(message = " 新密码不能为空 ")
    private String newPwd;

    /**
     * 确认密码
     */
    @NotBlank(message = " 确认密码不能为空 ")
    private String confirmPwd;

}
```

8. 返回报文及示例

参见公共返回示例。

20.6 用户信息修改接口

1. 接口用途

此接口用于用户修改昵称、更换头像。

2. 参数校验

昵称与用户头像不能同时为空；用户令牌不能为空。

3. 业务规则

用户令牌有效性校验。

4. 请求方式

采用 POST 请求。

5. 请求报文

用户信息修改接口请求报文如表 20-7 所示。

表 20-7　用户信息修改接口请求报文说明

参数名	必需	类型	字节长度	说明
userName	否	string	20	用户昵称
userAvatar	否	file	——	用户头像

6. 请求示例

参照注册接口。

7. 编写程序

（1）UserController 类，updateUserInfo 方法主要用于接收客户端修改用户信息请求。主要包含校验用户令牌是否为空、昵称与头像不能同时为空，然后调用 service 层修改用户信息服务，并返回报文给客户端。

```
/**
 * 修改用户信息
 * @return
 */
@RequestMapping(value = "updateUserInfo", method = RequestMethod.POST)
public ResponseBO updateUserInfo(UserInfoVO vo) {
    // 校验 token 是否为空
    String token = checkToken();
    boolean flag = !Detect.notEmpty(vo.getUserName()) && vo.getUserAvatar() == null;
    // 昵称与头像不能同时为空
    Assertion.isTrue(!flag, "参数异常");
    vo.setToken(token);
    // 调用 service 层业务修改用户信息
    userService.updateUserInfo(vo);
```

```
        return ResponseBO.success();
    }
```

（2）UserServiceImpl 类，updateUserInfo 方法用于修改用户信息业务。首先校验 token 是否有效，然后通过昵称及用户编号查询当前昵称是否已存在，若存在，抛出相应的异常信息，否则判断是否上传了头像；若 file 不为空，则调用 OSS 上传文件服务，然后判断是否修改了昵称；若昵称不为空，也设置到 update 实体中，然后统一将最新的用户信息同步至数据库及 redis 缓存。

```
/**
 * 修改用户信息
 * @param vo
 */
@Override
public void updateUserInfo(UserInfoVO vo) {
    User user = this.checkToken(vo.getToken());
    // 判断用户昵称是否重复
    Example example = new Example(User.class);
    example.createCriteria().andEqualTo("userName", vo.getUserName())
            .andNotEqualTo("userId", user.getUserId());
    User u = userMapper.selectOneByExample(example);
    Assertion.isNull(u, " 昵称已存在 ");
    User update = new User();
    MultipartFile file = vo.getUserAvatar();
    if (null != file) {
        // 上传头像，OSS 存储
        OSSClient ossClient = OssUtil.getOssClient();
        String bucketName = OssUtil.getBucketName();
        String diskName = "images/user/" + DateUtils.getTimeString(new Date());
        String fileName = System.currentTimeMillis() + "_" + file.getOriginalFilename();
        try {
            OssUtil.uploadInputStreamObject2Oss(ossClient, file.getInputStream(), fileName, bucketName, diskName);
        } catch (Exception e) {
            Assertion.isTrue(true, " 头像上传失败 ");
        }
```

```java
            String userAvatar = OssUtil.getOssUrl() + "/" + diskName + 
"/" + fileName;
            // 更新用户头像
            if (Detect.notEmpty(userAvatar)) {
                update.setUserAvatar(userAvatar);
            }
        }
        // 更新昵称
        if (Detect.notEmpty(vo.getUserName())) {
            update.setUserName(vo.getUserName());
        }
        update.setUserId(user.getUserId());
        BeanUtil.copyPropertiesIgnoreNull(update, user);
// 修改用户信息
this.updateUserInfo(update, user);
    }
```

（3）入参实体 UserInfoVO，参数为昵称及用户头像，不进行强校验，但是不能同时为空。

```java
package com.xpwu.secondary.vo;

import lombok.Getter;
import lombok.Setter;
import lombok.ToString;
import org.springframework.web.multipart.MultipartFile;

/**
 * Created by IntelliJ IDEA.
 *
 * @author:caoxue
 * @date:2019/8/13 16:43
 * @description:修改用户信息入参
 * @version:1.0
 */
@Setter
@Getter
@ToString(callSuper = true)
public class UserInfoVO extends BaseVO {

    private static final long serialVersionUID = 5372633025027598083L;
```

```
    /**
     * 昵称
     */
    private String userName;

    /**
     * 用户头像
     */
    private MultipartFile userAvatar;

}
```

8. 返回报文及示例

参见公共返回示例。

20.7 校验是否登录接口

1. 接口用途

此接口用于检查用户当前登录是否有效。

2. 参数校验

校验 token 是否为空。

3. 业务规则

校验 token 是否有效，若无效，直接抛出异常信息，code 返回值为 999。

4. 请求方式

采用 POST 请求。

5. 请求报文

公共请求报文，无额外报文。

6. 请求示例

参见公共请求示例。

7. 编写程序

（1）UserController 类，checkLoginValid 方法用于校验当前登录是否有效。主要包含校验 token 是否为空，然后调用 service 层退出登录方法，并返回报文信息给客户端。

```java
/**
 * 判断登录是否有效
 * @return
 */
@RequestMapping(value = "checkLoginValid", method = RequestMethod.POST)
public ResponseBO checkLoginValid() {
    // 校验token是否为空
    String token = checkToken();
    userService.checkToken(token);
    return ResponseBO.success();
}
```

（2）UserServiceImpl 类，checkToken 方法用于校验 token 有效性。通过 token 查询 redis 内存储的 token 信息，并根据 token 信息里面的手机号查询 redis 中的用户信息，若信息都存在，则证明 token 有效，返回客户端成功信息；若 redis 中未查询到相关信息，则根据 token 查询数据库中是否存在相应的用户数据；若存在，则将 token 信息同步至 redis，并返回成功信息；若 redis 跟数据库中均不存在相应信息，则说明 token 无效，返回 999 错误码。

```java
/**
 * 校验是否登录
 * @param token
 * @return
 */
@Override
public User checkToken(String token) {
    if (!Detect.notEmpty(token)) {
        // 无效登录专属异常 code 999，客户端用于跳转登录页面判断
        throw new BusinessException(CodeEnum.TOKEN_FAILURE);
    }
    User user;
    // 查询token是否有效
    TokenBO bo = new TokenBO();
    // 先查询redis
    String tokenInfo = RedisUtils.get(String.format(RedisConstants.TOKEN, token));
    if (Detect.notEmpty(tokenInfo)) {
        bo = JSONObject.parseObject(tokenInfo, TokenBO.class);
```

```java
            // 查询用户信息
            String userInfo = RedisUtils.get(String.format(RedisConstants.
USER_INFO, bo.getMobile()));
            if (!Detect.notEmpty(userInfo)) {
                user = this.findUserByToken(token);
                RedisUtils.setEx(String.format(RedisConstants.USER_INFO,
bo.getMobile()), RedisConstants.USER_INFO_EXPIRE, JSON.toJSONString(user));
            } else {
                user = JSONObject.parseObject(userInfo, User.class);
            }
        } else {
            // redis 无信息，再查询数据库
            user = this.findUserByToken(token);
            bo.setMobile(user.getMobile());
            bo.setUserId(user.getUserId());
            int expired = (int) ((user.getTokenExpired() - System.
currentTimeMillis()) / 1000);
            RedisUtils.setEx(String.format(RedisConstants.TOKEN, token),
expired, JSON.toJSONString(bo));
        }
        Assertion.notNull(user, "用户异常");
        return user;
    }
```

8. 返回报文及示例

参见公共返回示例。

20.8　退出登录接口

1. 接口用途

此接口用于用户退出当前系统。

2. 参数校验

校验 token 是否为空。

3. 业务规则

校验 token 是否有效，若无效，直接抛出异常信息；若有效，修改用户状态为未登录，删除 redis 里面的 token 信息及用户信息，完成退出操作。

4. 请求方式

采用 POST 请求。

5. 请求报文

公共请求报文，无额外报文。

6. 请求示例

参见公共请求示例。

7. 编写程序

（1）UserController 类，logout 方法用于接收客户端退出登录请求。主要包含校验 token 是否为空，然后调用 service 层退出登录业务，并返回报文信息给客户端。

```java
/**
 * 退出登录
 * @return
 */
@RequestMapping(value = "logout", method = RequestMethod.POST)
public ResponseBO logout() {
    // 校验token是否为空
    String token = checkToken();
    userService.logout(token);
    return ResponseBO.success();
}
```

（2）UserServiceImpl 类，logout 方法用于退出登录业务。校验 token 是否有效，将传入的 token 作为 key 查询 redis 信息，若查不到信息，说明 token 无效。否则修改数据库用户登录状态为未登录，同时删除 redis 内 token 信息及用户信息。

```java
/**
 * 退出登录
 * @param token
 */
@Override
public void logout(String token) {
    String info = RedisUtils.get(String.format(RedisConstants.TOKEN, token));
    if (!Detect.notEmpty(info)) {
        // token失效
```

```
            throw new BusinessException(CodeEnum.TOKEN_FAILURE);
        }
        TokenBO bo = JSONObject.parseObject(info, TokenBO.class);
        User user = new User();
        user.setUserId(bo.getUserId());
        user.setLoginStatus(2);
        user.setUpdateTime(new Date());
        userMapper.updateByPrimaryKeySelective(user);
        // 删除 redis token 信息
        RedisUtils.del(String.format(RedisConstants.TOKEN, token));
        // 删除 redis 用户信息
        RedisUtils.del(String.format(RedisConstants.USER_INFO, bo.getMobile()));
    }
```

8. 返回报文及示例

参见公共返回示例。

第 21 章 商品模块接口

【本章导读】

◎ 商品类别列表接口

◎ 商品列表接口

◎ 发布商品接口

◎ 修改商品接口

◎ 获取商品详情接口

◎ 评论 / 回复接口

◎ 获取评论 / 回复列表接口

◎ 点赞 / 取消赞接口

◎ 获取点赞列表接口

◎ 首页轮播商品列表接口

21.1 商品类别列表接口

1. 接口用途

此接口用于获取商品类别列表，下拉列表框展示，用于用户发布 / 修改商品。

2. 参数校验

无参查询。

3. 业务规则

查询商品类别表有效数据。

4. 请求方式

采用 GET 请求。

5. 请求报文

无。

6. 请求示例

无。

7. 编写程序

（1）ProductTypeController 类，getProductTypeList 方法用于接收客户端查询商品类别列表请求。直接调用 service 层业务，查询有效数据，并返回报文给客户端。

```java
/**
 * 获取商品类别列表
 * @return
 */
@RequestMapping(value = "getProductTypeList", method = RequestMethod.GET)
public ResponseBO getProductTypeList() {
    Example example = new Example(ProductType.class);
    Example.Criteria criteria = example.createCriteria();
    // 查询有效数据
    criteria.andEqualTo("status", 1);
    return ResponseBO.success(productTypeService.findList(example));
}
```

（2）service 层，此处采用基础 service 封装查询，同时 ProductTypeService 继承 BaseService 方式调用 findList 方法。

```java
/**
 * 基础 service
 * @author caoxue
 * @version 1.0
 * @date 2019/8/6 10:20
 */
public interface BaseService<T, D> {
    /**
     * 保存
     * @param entity
     * @return
     * @throws BusinessException
     */
    int save(T entity) throws BusinessException;
```

```java
    /**
     * 修改
     * @param entity
     * @return
     * @throws BusinessException
     */
    int update(T entity) throws BusinessException;

    /**
     *
     * @param id
     * @return
     * @throws BusinessException
     */
    int delete(D id) throws BusinessException;

    /**
     * 根据ID查找
     * @param id
     * @return
     */
    T findById(D id);

    /**
     * 列表查询
     * @param example
     * @return
     */
    List<T> findList(Example example);
```
ProductTypeService 类
```java
    public interface ProductTypeService extends BaseService<ProductType, Integer> {
    }
```

（3）BaseServiceImpl 基类。

```java
/**
 * 基类的实现
 *
 * @author caoxue
```

```java
 * @version 1.0
 * @date 2019/8/6 10:21
 */
public class BaseServiceImpl<T, D> implements BaseService<T, D> {

    @Autowired
    public Mapper<T> mapper;

    @Override
    public int save(T entity) throws BusinessException {
        return mapper.insertSelective(entity);
    }

    @Override
    public int update(T entity) throws BusinessException {
        return mapper.updateByPrimaryKeySelective(entity);
    }

    @Override
    public int delete(D id) throws BusinessException {
        return mapper.deleteByPrimaryKey(id);
    }

    @Override
    public T findById(D id) {
        return mapper.selectByPrimaryKey(id);
    }

    @Override
    public List<T> findList(Example example) {
        return mapper.selectByExample(example);
    }

    public Mapper<T> getMapper() {
        return mapper;
    }

    public void setMapper(Mapper<T> mapper) {
        this.mapper = mapper;
    }
}
```

ProductTypeServiceImpl 类继承 BaseServiceImpl 基类，并调用父类 findList 方法，主要通过 example 查询 status=1 的数据，即有效数据。

```
@Service
public class ProductTypeServiceImpl extends BaseServiceImpl<Product
Type, Integer> implements ProductTypeService {
}
```

8. 返回报文

商品类别列表接口返回报文如表 21-1 所示。

表 21-1 商品类别列表接口返回报文说明

参数名	必需	类型	字节长度	说明
id	是	int	11	商品类别编号
name	是	string	100	商品类别名称

9. 返回示例

```
{
"code":0,
"message":"success",
"requestId":null,
"timestamp":1567301655000,
"result":[
{
"id":1,
"name":"手机"
},
{
"id":2,
"name":"母婴"
}
]
}
```

21.2　商品列表接口

1. 接口用途

此接口用于首页商品列表查询，可以根据商品名称模糊搜索、商品类别精确搜索。

2. 参数校验

动态参数查询，商品描述、商品类别参数均可以不传，当前页码不传默认为1，每页条数不传默认为10。

3. 业务规则

交易中的数据不展示在首页，未交易的数据展示在前面，已完成交易的数据展示在后面。

4. 请求方式

采用 POST 请求。

5. 请求报文

商品列表接口请求报文如表 21-2 所示。

表 21-2　商品列表接口请求报文说明

参数名	必需	类型	字节长度	说明
productDesc	否	string	100	商品描述
productTypeId	否	int	6	商品类别编号
pageNum	否	int	6	当前页码，不传默认为1
pageSize	否	int	6	每页条数，不传默认为10

6. 请求示例

```
{
"productDesc":"手机",
"productTypeId":1,
"pageNum":1,
"pageSize":10
}
```

7. 编写程序

（1）ProductController 类为商品相关接口的控制层，getProductList 方法用于接收

客户端首页商品信息查询请求。通过 PageHelper 工具开启分页查询，并调用 service 层业务，将报文返回给客户端。

```java
    /**
     * 获取商品列表，不需要登录
     * @param vo
     * 商品名称，商品类别动态查询
     * @return
     */
    @RequestMapping(value = "getProductList", method = RequestMethod.POST)
    public ResponseBO getProductList(@RequestBody ProductSearchVO vo) {
        // 使用 PageHelper 工具直接开启分页查询
        PageHelper.startPage(vo.getPageNum(), vo.getPageSize());
        // 调用 service 层查询方法并返回给客户端
        return ResponseBO.successPageInfo(productService.getProductList(vo));
    }

    /**
     * 分页数据封装
     *
     * @param list
     * 数据列表
     * @return
     */
    public static ResponseBO successPageInfo(List<?> list) {
        Map<String, Object> map = new HashMap<>(16);
        map.put("list", list);
        map.put("total", Math.toIntExact(new PageInfo<>(list).getTotal()));
        return new ResponseBO.Builder().result(map).build();
    }
```

（2）ProductServiceImpl 类，getProductList 方法用于首页商品列表查询业务，调用 Mapper 层查询。

```java
    /**
     * 查询首页商品列表
```

```
 * @param vo
 * @return
 */
@Override
public List<ProductBO> getProductList(ProductSearchVO vo) {
    return productMapper.selectProductList(vo);
}
```

（3）ProductMapper 类，自定义 selectProductList 接口查询。

```
/**
 * 查询首页商品列表
 * @param vo
 * @return
 */
List<ProductBO> selectProductList(ProductSearchVO vo);
```

（4）ProductMapper.xml，此处属于扩展查询，需要另外写 SQL 语句连接，具体如下。

```
<!-- 首页商品查询 -->
<select id="selectProductList" parameterType="com.xpwu.secondary.vo.ProductSearchVO" resultType="com.xpwu.secondary.bo.ProductBO">
    SELECT
        p.id,
        p.product_imgs productImgs,
        p.product_desc productDesc,
        p.product_price productPrice,
        p.product_type_id productTypeId,
        t.wantNum,
        pt.'name' productTypeName,
        p.publish_user_id publishUserId,
        u.user_name publishUserName,
        u.user_avatar publishUserAvatar,
        o.buying_user_id buyingUserId,
        o.trade_status tradeStatus,
        p.create_time createTime,
        o.completion_time tradeTime,
```

```
            p.product_address productAddress
    FROM
        product p
    LEFT JOIN product_type pt ON p.product_type_id = pt.id
    LEFT JOIN 'order' o ON p.id = o.product_id
    LEFT JOIN (SELECT COUNT(1) wantNum,product_id FROM chat c GROUP BY
product_id) t ON p.id = t.product_id
    LEFT JOIN 'user' u ON p.publish_user_id = u.user_id
    <where>
        <if test="productDesc != null and productDesc != ''">
            and p.product_desc like concat('%', #{productDesc}, '%')
        </if>
        <if test="productTypeId != null">
            and p.product_type_id = #{productTypeId}
        </if>
        and (o.order_id is null or o.trade_status in (1, 2))
    </where>
    group by p.id
    order by IFNULL(o.trade_status, 4) DESC, p.create_time DESC
</select>
```

（5）入参实体 ProductSearchVO，对应客户端请求报文。

```
package com.xpwu.secondary.vo;

import lombok.Getter;
import lombok.Setter;
import lombok.ToString;

/**
 * Created by IntelliJ IDEA.
 *
 * @author:caoxue
 * @date:2019/8/8 16:42
 * @description: 首页商品查询接口入参
 * @version:1.0
 */
@Setter
@Getter
@ToString(callSuper = true)
```

```java
public class ProductSearchVO extends BaseVO {
    private static final long serialVersionUID = 1754850472175512237L;

    /**
     * 商品描述
     */
    private String productDesc;

    /**
     * 商品类别编号
     */
    private Integer productTypeId;

}
```

(6) 返回报文实体 ProductBO。

```java
package com.xpwu.secondary.bo;

import lombok.Getter;
import lombok.Setter;
import lombok.ToString;

import java.io.Serializable;
import java.math.BigDecimal;
import java.util.Date;

/**
 * Created by IntelliJ IDEA.
 *
 * @author:caoxue
 * @date:2019/9/12 9:55
 * @description:
 * @version:1.0
 */
@Setter
@Getter
@ToString(callSuper = true)
public class ProductBO implements Serializable {
```

```java
    private static final long serialVersionUID = 8821081481707766726L;

    /**
     * 商品编号
     */
    private Integer id;

    /**
     * 商品图片,多个图片地址用英文逗号隔开
     */
    private String productImgs;

    /**
     * 商品描述
     */
    private String productDesc;

    /**
     * 商品价格
     */
    private BigDecimal productPrice;

    /**
     * 商品类别编号
     */
    private Integer productTypeId;

    /**
     * 商品类别名称
     */
    private String productTypeName;

    /**
     * 商品发布人用户编号
     */
    private String publishUserId;

    /**
     * 商品发布人用户昵称
     */
    private String publishUserName;
```

```java
    /**
     * 商品发布人用户头像
     */
    private String publishUserAvatar;

    /**
     * 购买人用户编号
     */
    private String buyingUserId;

    /**
     * 交易状态: 1——未交易, 2——交易中, 3——交易成功, 4——交易失败
     */
    private Integer tradeStatus;

    /**
     * 创建时间 (发布时间)
     */
    private Date createTime;

    /**
     * 交易时间 (购买时间)
     */
    private Date tradeTime;

    /**
     * 想要人数
     */
    private Integer wantNum;

    /**
     * 订单编号
     */
    private Integer orderId;

    /**
     * 商品地址
     */
    private String productAddress;

}
```

8. 返回报文

商品列表接口返回报文如表 21-3 所示。

表 21-3 商品列表接口返回报文说明

参数名	必需	类型	字节长度	说明
id	是	int	11	商品编号
productImgs	是	string	2000	商品图片，多个图片地址用英文逗号隔开
productDesc	是	string	500	商品描述
productPrice	是	double	10	商品价格
productTypeId	是	Int	11	商品类别编号
productTypeName	是	string	100	商品类别名称
publishUserId	是	string	32	商品发布人用户编号
publishUserName	是	string	50	商品发布人用户昵称
publishUserAvatar	是	string	100	商品发布人用户头像
buyingUserId	否	string	32	购买人用户编号
createTime	是	long	13	创建时间（发布时间）
wantNum	是	int	11	想要人数（有多少人想要此商品）
tradeTime	否	long	13	交易时间（购买时间）
productAddress	是	string	500	商品地址

9. 返回示例

```
{
"code":0,
"message":"success",
"requestId":null,
"timestamp":1567301655000,
"result":[
{
"id":1,
"productImgs":"xxx",
"productDesc":"全新手机",
"productPrice":4000,
"productTypeId":1,
"productTypeName":"手机",
"publishUserId":"xxx",
```

```
"publishUserName":"xxx",
"publishUserAvatar":"xxx",
"buyingUserId":"xxx",
"createTime":1567301655000,
"wantNum":5,
"tradeTime":null,
"productAddress":"xxx"
        }
    ]
}
```

21.3 发布商品接口

1. 接口用途

此接口用于用户发布商品信息，主要信息包括商品图片、商品描述、商品价格、商品类别编号、商品地址。通过程序校验并写入商品表中。

2. 参数校验

校验各参数是否为空，参数包括用户令牌、商品价格、商品类别编号、商品地址及商品图片校验。

3. 业务规则

校验用户令牌是否有效，确认用户是登录后进行操作。上传用户头像，并将商品信息息入库。

4. 请求方式

采用 POST 请求。

5. 请求报文

发布商品接口请求报文如表 21-4 所示。

表 21-4 发布商品接口请求报文说明

参数名	必需	类型	字节长度	说明
productImgs	是	file	——	商品图片，支持多张图片
productPrice	是	double	10	商品价格
productTypeId	是	int	11	商品类别编号
productAddress	是	string	500	商品地址
productDesc	否	string	500	商品描述

6. 请求示例

使用 Postman 工具，参见用户注册示例。

7. 编写程序

（1）ProductController 类，商品相关的接口控制层，publishProduct 方法用于接收客户端发布商品请求。主要包含对参数进行基础校验，然后调用 service 层发布商品业务，并将报文返回给客户端。

```java
/**
 * 发布商品
 * @param productImgs
 * @param vo
 * @return
 */
@RequestMapping(value = "publishProduct")
public ResponseBO publishProduct(@RequestParam("productImgs")
MultipartFile[] productImgs, @Valid PublishProductVO vo) {
    // 校验token是否为空
    String token = checkToken();
    Assertion.notNull(productImgs, "请选择图片");
    vo.setToken(token);
    // 调用发布商品业务
    productService.publishProduct(productImgs, vo);
    return ResponseBO.success();
}
```

（2）ProductServiceImpl 类，publishProduct 方法用于发布商品信息业务。首先校验 token 是否有效，确认用户是登录后进行操作，然后对图片进行上传，若上传失败，直接抛出对应的异常信息。否则将商品信息组装之后插入商品表中，完成发布商品信息业务。

```java
/**
 * 发布商品
 * @param files
 * @param vo
 */
@Override
public void publishProduct(MultipartFile[] files, PublishProductVO vo) {
```

```java
        User user = userService.checkToken(vo.getToken());
        // OSS 存储
        OSSClient ossClient = OssUtil.getOssClient();
        String bucketName = OssUtil.getBucketName();
        String diskName = "images/product/" + DateUtils.getTimeString(new Date());
        StringBuilder stringBuilder = null;
        try {
            stringBuilder = OssUtil.batchUploadInputStreamObject2Oss(ossClient, files, bucketName, diskName);
        } catch (Exception e) {
            Assertion.isTrue(true, "上传失败");
        }
        Assertion.notNull(stringBuilder, "文件上传失败");
        String productImgs = stringBuilder.substring(0, stringBuilder.length() - 1);
        Product product = new Product();
        Date date = new Date();
        BeanUtils.copyProperties(vo, product);
        product.setProductImgs(productImgs);
        product.setCreateTime(date);
        product.setUpdateTime(date);
        product.setPublishUserId(user.getUserId());
        productMapper.insertSelective(product);
    }
```

（3）OssUtil 类，阿里云 OSS 文件上传工具方法，上传多个文件。

```
/**
 * 向阿里云的 OSS 存储中存储文件，批量上传
 *
 * @param client
 * OSS 客户端
 * @param files
 * 上传文件
 * @param bucketName
 * bucket 名称
 * @param diskName
 * 上传文件的目录，bucket 下文件的路径
 * @return String 唯一 MD5 数字签名
```

```java
     */
    public static StringBuilder batchUploadInputStreamObject2Oss(OSSClient 
client, MultipartFile[] files, String bucketName, String diskName) {
        StringBuilder builder = new StringBuilder();
        try {
            log.info("upload start");
            // 创建上传Object的Metadata
            for (MultipartFile file :files) {
                try {
                    String fileName = System.currentTimeMillis() + "_" + 
file.getOriginalFilename();
                    fileName = fileName.trim();
                    InputStream is = file.getInputStream();
                        ObjectMetadata metadata = new ObjectMetadata();
                        metadata.setContentLength(is.available());
                        metadata.setCacheControl("no-cache");
                        metadata.setHeader("Pragma", "no-cache");
                        metadata.setContentEncoding("UTF-8");
                        metadata.setContentType(getContentType(fileName));
                        log.info("uploading" + bucketName + "/" + diskNa
me + "/" + fileName);
                    // 上传文件
                        PutObjectResult ddd = client.putObject(bucketNam
e, diskName + "/" + fileName, is, metadata);
                        log.info("uploaded");
                        log.info("上传文件 result :" + JSONObject.toJSONString
(ddd));
                        builder.append(OssUtil.getOssUrl()).append("/").
append(diskName).append("/").append(fileName).append(",");
                        is.close();
                        log.info("关闭文件的输入流！");
                } catch (Exception e) {
                    log.error("上传文件到OSS失败", e);
                    return null;
                }

            }

        } catch (Exception e) {
            log.error("上传文件到OSS失败", e);
            return null;
```

```
        } finally {
            client.shutdown();
        }
        return builder;
    }
```

（4）入参实体 PublishProductVO，对应客户端部分请求报文。

```
package com.xpwu.secondary.vo;

import lombok.Getter;
import lombok.Setter;
import lombok.ToString;

import javax.validation.constraints.NotBlank;
import javax.validation.constraints.NotNull;
import java.math.BigDecimal;

/**
 * Created by IntelliJ IDEA.
 *
 * @author:caoxue
 * @date:2019/8/10 11:22
 * @description:发布商品接口入参
 * @version:1.0
 */
@Setter
@Getter
@ToString(callSuper = true)
public class PublishProductVO extends BaseVO {
    private static final long serialVersionUID = -3481201029921085411L;

    /**
     * 商品描述
     */
    private String productDesc;

    /**
     * 商品价格
     */
```

```
    @NotNull(message = "商品价格不能为空")
    private BigDecimal productPrice;

    /**
     * 商品类别编号
     */
    @NotNull(message = "商品类别不能为空")
    private Integer productTypeId;

    /**
     * 商品地址
     */
    @NotBlank(message = "商品地址不能为空")
    private String productAddress;

}
```

8. 返回报文及示例

参见公共返回示例。

21.4 修改商品信息接口

1. 接口用途

此接口用于用户修改自己发布的商品信息，商品信息主要包括商品图片、商品描述、商品价格、商品地址、商品类别编号、商品编号。

2. 参数校验

校验各参数是否为空，参数包括用户令牌、商品编号。

3. 业务规则

校验用户令牌是否有效，根据商品编号查询商品信息是否存在，确认商品发布人是否为当前登录用户，修改商品信息并入库。

4. 请求方式

采用 POST 请求。

5. 请求报文

修改商品信息接口请求报文如表 21-5 所示。

表 21-5　修改商品信息接口请求报文说明

参数名	必需	类型	字节长度	说明
productId	是	int	11	商品编号
productImgs	是	file	——	商品图片，支持多张
productPrice	否	double	10	商品价格
productTypeId	否	int	11	商品类别编号
productAddress	否	string	500	商品地址
productDesc	否	string	500	商品描述
oldImgs	否	string	2000	未变更的图片地址，多个地址用英文逗号隔开

6. 请求示例

使用 Postman 工具，参见用户注册示例。

7. 编写程序

（1）ProductController 类，商品相关的接口控制层，updateProduct 方法用于接收客户端修改商品信息请求。主要包含对参数进行基础校验，然后调用 service 层修改商品业务，并将报文返回给客户端。

```java
/**
 * 修改商品信息
 * @param productImgs
 * @param vo
 * @return
 */
@RequestMapping(value = "updateProduct", method = RequestMethod.POST)
public ResponseBO updateProduct(@RequestParam("productImgs")
MultipartFile[] productImgs, @Valid UpdateProductVO vo) {
    // 校验token是否为空
    String token = checkToken();
    vo.setToken(token);
    // 调用修改商品service
    productService.updateProduct(productImgs, vo);
    return ResponseBO.success();
}
```

（2）ProductServiceImpl 类，updateProduct 方法用于修改商品信息业务。首先校

验 token 是否有效，确认用户是登录后进行操作，然后根据商品编号查询商品是否存在，若不存在，直接抛出对应的异常信息。否则继续确认该商品发布人是否为当前登录人，只有商品发布人自己可以修改自己发布的商品信息，修改后将商品信息更新到商品表中，完成修改商品信息业务。

```java
/**
 * 修改商品信息
 * @param files
 * @param vo
 */
@Override
public void updateProduct(MultipartFile[] files, UpdateProductVO vo) {
    // 校验 token 是否有效
    User user = userService.checkToken(vo.getToken());
    // 查询商品是否存在
    Product product = productMapper.selectByPrimaryKey(vo.getProductId());
    Assertion.notNull(product, "商品信息不存在");
    // 确认商品发布人为当前登录人
    Assertion.equals(product.getPublishUserId(), user.getUserId(), "只有商品发布人可以修改");
    String productImgs = null;
    StringBuilder stringBuilder = new StringBuilder();
    // 修改商品图片
    if (Detect.notEmpty(files)) {
        // OSS 存储
        OSSClient ossClient = OssUtil.getOssClient();
        String bucketName = OssUtil.getBucketName();
        String diskName = "images/product/" + DateUtils.getTimeString(new Date());
        try {
            // 批量上传图片
            stringBuilder = OssUtil.batchUploadInputStreamObject2Oss(ossClient, files, bucketName, diskName);
        } catch (Exception e) {
            Assertion.isTrue(true, "上传失败");
        }
        Assertion.notNull(stringBuilder, "文件上传失败");
    }
```

```java
        if (Detect.notEmpty(vo.getOldImgs())) {
            for (String url :vo.getOldImgs()) {
                stringBuilder.append(url).append(",");
            }
        }
        if (stringBuilder.length() > 1) {
            productImgs = stringBuilder.substring(0, stringBuilder.length() - 1);
        }
        // 修改用户信息
        Product update = new Product();
        Date date = new Date();
        BeanUtils.copyProperties(vo, update);
        update.setProductImgs(productImgs);
        update.setId(vo.getProductId());
        update.setUpdateTime(date);
        productMapper.updateByPrimaryKeySelective(update);
    }
```

（3）入参实体 UpdateProductVO，对应客户端部分请求报文。

```java
package com.xpwu.secondary.vo;

import lombok.Getter;
import lombok.Setter;
import lombok.ToString;

import javax.validation.constraints.NotNull;
import java.math.BigDecimal;

/**
 * Created by IntelliJ IDEA.
 *
 * @author:caoxue
 * @date:2019/9/13 15:32
 * @description:修改商品信息入参
 * @version:1.0
 */
@Setter
@Getter
```

```java
@ToString(callSuper = true)
public class UpdateProductVO extends BaseVO {

    private static final long serialVersionUID = 7193929719742570346L;

    /**
     * 商品描述
     */
    private String productDesc;

    /**
     * 商品价格
     */
    private BigDecimal productPrice;

    /**
     * 商品类别编号
     */
    private Integer productTypeId;

    /**
     * 商品编号
     */
    @NotNull(message = "商品编号不能为空")
    private Integer productId;

    /**
     * 未变更的图片地址
     */
    private String[] oldImgs;

    /**
     * 商品地址
     */
    private String productAddress;

}
```

8. 返回报文及示例

参见公共返回示例。

21.5 获取商品详情接口

1. 接口用途

此接口用于获取商品详情,可查看单个商品详细信息。

2. 参数校验

校验商品编号是否为空。

3. 业务规则

关联查询出商品有多少人想要及点赞数据。

4. 请求方式

采用 GET 请求。

5. 请求报文

获取商品详情接口请求报文如表 21-6 所示。

表 21-6 获取商品详情接口请求报文说明

参数名	必需	类型	字节长度	说明
productId	否	int	11	商品编号

6. 请求示例

GET 请求,将 productId 放在接口 URL 后面即可。

7. 编写程序

(1) ProductController 类为商品相关接口的控制层,getProductDetail 方法用于商品详情信息查询。主要包含校验请求参数,并调用 service 层业务,将报文返回给客户端。

```
/**
 * 获取商品详情
 * @param productId
 * @return
 */
@RequestMapping(value = "getProductDetail", method = RequestMethod.GET)
public ResponseBO getProductDetail(Integer productId) {
    // 校验商品编号
    Assertion.isPositive(productId, "商品编号不能为空");
    // 调用 service 层业务
    return ResponseBO.success(productService.getProductDetail
```

```
(productId));
    }
```

（2）ProductServiceImpl 类，getProductDetail 方法用于商品详情信息查询业务，直接调用 Mapper 层查询。

```java
/**
 * 获取商品详情
 * @param productId
 * @return
 */
@Override
public ProductDetailsBO getProductDetail(Integer productId) {
    return productMapper.selectProductDetail(productId);
}
```

（3）ProductMapper.xml，此处属于扩展查询，需要另外写 SQL 语句，具体如下。

```xml
<!-- 商品详情 -->
<select id="selectProductDetail" parameterType="integer" resultType="com.xpwu.secondary.bo.ProductDetailsBO">
    SELECT
        pro.publish_user_id publishUserId,
        us.user_name publishUserName,
        us.user_avatar publishUserAvatar,
        pro.create_time publishTime,
        IFNULL(pr.'status',2) praiseStatus,
        pro.id productId,
        pro.product_imgs productImgs,
        pro.product_desc productDesc,
        pro.product_price productPrice,
        pt.id productTypeId,
        pt.'name' productTypeName,
        t.wantNum,
        pro.product_address productAddress

    FROM product pro
    LEFT JOIN 'user' us ON pro.publish_user_id = us.user_id
    LEFT JOIN (SELECT COUNT(1) wantNum,product_id FROM chat c GROUP
```

```
BY product_id) t ON pro.id = t.product_id
    LEFT JOIN product_type pt ON pro.product_type_id = pt.id
    LEFT JOIN praise pr ON pro.id = pr.product_id AND pr.user_id =
us.user_id
    WHERE pro.id = #{productId}

</select>
```

(4)出参实体 ProductDetailsBO，响应报文实体。

```
package com.xpwu.secondary.bo;

import lombok.Getter;
import lombok.Setter;
import lombok.ToString;

import java.io.Serializable;
import java.math.BigDecimal;
import java.util.Date;

/**
 * Created by IntelliJ IDEA.
 *
 * @author:caoxue
 * @date:2019/8/18 0:19
 * @description:
 * @version:1.0
 */
@Setter
@Getter
@ToString(callSuper = true)
public class ProductDetailsBO implements Serializable {

    private static final long serialVersionUID = 2808501232203226702L;

    /**
     * 商品发布人用户编号
     */
    private String publishUserId;

    /**
```

* 商品发布人用户昵称
 */
private String publishUserName;

/**
 * 商品发布人用户头像
 */
private String publishUserAvatar;

/**
 * 商品发布时间
 */
private Date publishTime;

/**
 * 点赞状态：1——点赞，2——未点赞
 */
private Integer praiseStatus;

/**
 * 商品编号
 */
private Integer productId;

/**
 * 商品图片
 */
private String productImgs;

/**
 * 商品描述
 */
private String productDesc;

/**
 * 商品价格
 */
private BigDecimal productPrice;

/**
 * 商品类别编号
 */

```java
    private Integer productTypeId;

    /**
     * 商品类别名称
     */
    private String productTypeName;

    /**
     * 想要人数
     */
    private Integer wantNum;

    /**
     * 商品地址
     */
    private String productAddress;

}
```

8. 返回报文

获取商品详情接口返回报文如表 21-7 所示。

表 21-7 获取商品详情接口返回报文说明

参数名	必需	类型	字节长度	说明
publishUserId	是	int	11	商品发布人用户编号
publishUserName	是	string	50	商品发布人用户昵称
publishUserAvatar	是	string	100	商品发布人用户头像
publishTime	是	long	13	商品发布时间
praiseStatus	是	int	2	点赞状态：1——点赞，2——未点赞
productId	是	int	11	商品编号
productImgs	是	string	2000	商品图片，多个图片地址用英文逗号隔开
productDesc	是	string	500	商品描述
productPrice	是	double	10	商品价格
productTypeId	是	Int	11	商品类别编号
productTypeName	是	string	100	商品类别名称
wantNum	是	int	11	想要人数（有多少人想要此商品）
productAddress	是	string	500	商品地址

9. 返回示例

返回示例如下。

```
{
"code":0,
"message":"success",
"requestId":null,
"timestamp":1567301655000,
"result":{
"productImgs":"xxx",
"productId":1,
"productDesc":"全新手机",
"productPrice":4000,
"productTypeId":1,
"productTypeName":"手机",
"publishUserId":"xxx",
"publishUserName":"xxx",
"publishUserAvatar":"xxx",
"praiseStatus":1,
"publishTime":1567301655000,
"wantNum":5,
"productAddress":"xxx"
}
}
```

21.6 评论/回复接口

1. 接口用途

此接口用于用户对商品进行评论或者回复评论。

2. 参数校验

校验用户令牌是否为空，校验商品编号是否为空，校验评论内容是否为空。

3. 业务规则

根据 token 查询用户是否存在，即 token 是否有效；根据商品编号查询商品是否存在；回复目标用户与回复编号要么同时传（回复），要么同时不传（评论）。

4. 请求方式

采用 POST 请求。

5. 请求报文

评论 / 回复接口请求报文如表 21-8 所示。

表 21-8　评论 / 回复接口请求报文说明

参数名	必需	类型	字节长度	说明
productId	是	int	11	商品编号
content	是	string	500	评论 / 回复内容
toUserId	否	string	32	回复目标用户（传参则表示回复，不传则表示评论）
replyId	否	int	11	回复编号（评论列表返回数据的主键 id，回复时必传）

6. 请求示例

```
{
"productId":1,
"content":"测试",
"toUserId":"xxx",
"replyId": 1
}
```

7. 编写程序

（1）CommentReplyController 类，评论 / 回复相关的接口控制层，commentOrReply 方法用于接收客户端评论 / 回复请求。主要包含校验用户令牌、商品编号、评论内容是否为空，然后调用 service 层评论 / 回复业务，并返回报文信息给客户端。

```
/**
 * 评论 / 回复接口
 * @param vo
 * @return
 */
@RequestMapping(value = "commentOrReply", method = RequestMethod.POST, produces = MediaType.APPLICATION_JSON_UTF8_VALUE)
public ResponseBO commentOrReply(@Valid @RequestBody CommentReplyVO vo) {
    // 校验 token 是否为空
    String token = checkToken();
    vo.setToken(token);
```

```
        // 调用 service 层评论 / 回复业务
        commentRelyService.commentOrReply(vo);
        return ResponseBO.success();
    }
```

（2）CommentReplyServiceImpl 类，commentOrReply 方法用于评论 / 回复业务。校验 token 有效性并返回用户实体，若校验不通过，则返回相应的错误信息。否则继续判断商品信息是否存在，若不存在，则返回相应的错误信息。否则通过 toUserId 判断是评论还是回复功能，若是评论，直接将相关信息组装入库；若是回复，则根据商品编号及回复目标用户查询是否有相关的评论 / 回复记录。若无相关记录，说明此回复参数异常，抛出对应的错误信息；若有记录，则将回复信息一起组装入库，完成评论 / 回复业务。

```
/**
 * 评论 / 回复业务
 * @param vo
 */
@Override
public void commentOrReply(CommentReplyVO vo) {
    // 校验 token 是否有效
    User user = userService.checkToken(vo.getToken());
    // 查询商品是否存在
    Product product = productMapper.selectByPrimaryKey(vo.getProductId());
    Assertion.notNull(product, "商品不存在");
    CommentReply commentReply = new CommentReply();
    // 回复
    if (Detect.notEmpty(vo.getToUserId())) {
        Assertion.isPositive(vo.getReplyId(), "replyId 不能为空");
        Example example = new Example(CommentReply.class);
        example.createCriteria().andEqualTo("productId", vo.getProductId())
                .andEqualTo("fromUserId", vo.getToUserId());
        List<CommentReply> list = commentReplyMapper.selectByExample(example);
        Assertion.notEmpty(list, "回复异常，未找到相关评论 / 回复记录");
        User toUser = userService.findById(vo.getToUserId());
        commentReply.setToUserAvatar(toUser.getUserAvatar());
```

```java
            commentReply.setToUserName(toUser.getUserName());
        } else {
            Assertion.isTrue(!Detect.isPositive(vo.getReplyId()), "参数异常：replyId");
        }
        BeanUtils.copyProperties(vo, commentReply);
        commentReply.setCreateTime(new Date());
        commentReply.setType(Detect.notEmpty(vo.getToUserId()) ? 2 :1);
        commentReply.setFromUserId(user.getUserId());
        commentReply.setFromUserAvatar(user.getUserAvatar());
        commentReply.setFromUserName(user.getUserName());
        commentReplyMapper.insertSelective(commentReply);
    }
```

（3）入参实体 CommentReplyVO，对应客户端请求报文。

```java
package com.xpwu.secondary.vo;

import lombok.Getter;
import lombok.Setter;
import lombok.ToString;

import javax.validation.constraints.NotBlank;
import javax.validation.constraints.NotNull;

/**
 * Created by IntelliJ IDEA.
 *
 * @author:caoxue
 * @date:2019/8/17 13:32
 * @description:评论 / 回复入参
 * @version:1.0
 */
@Setter
@Getter
@ToString(callSuper = true)
public class CommentReplyVO extends BaseVO {

    private static final long serialVersionUID = 6388061367719863036L;
```

```
/**
 * 商品编号
 */
@NotNull(message = "商品编号不能为空")
private Integer productId;

/**
 * 评论/回复内容
 */
@NotBlank(message = "内容不能为空")
private String content;

/**
 * 回复目标用户 传参则表示回复 不传则表示评论
 */
private String toUserId;

/**
 * 回复编号
 */
private Integer replyId;

}
```

8. 返回报文及示例

参见公共返回示例。

21.7 评论/回复列表接口

1. 接口用途

此接口用于查询商品相关的评论/回复列表数据。

2. 参数校验

校验商品编号是否为空。

3. 业务规则

根据评论/回复时间升序排序返回给客户端。

4. 请求方式

采用 GET 请求。

5. 请求报文

评论回复 / 列表接口请求报文如表 21-9 所示。

表 21-9 评论 / 回复列表接口请求报文说明

参数名	必需	类型	字节长度	说明
productId	否	int	11	商品编号

6. 请求示例

GET 请求，将 productId 放在接口 URL 后面即可。

7. 编写程序

（1）CommentReplyController 类为评论 / 回复相关的接口控制层，getCommentReplyList 方法用于查询评论 / 回复列表数据，并校验请求参数商品编号，然后组装通用查询，调用 service 层业务，将报文返回给客户端。

```
/**
 * 获取评论 / 回复列表
 * @param productId
 * @return
 */
@RequestMapping(value = "getCommentReplyList", method = RequestMethod.GET)
public ResponseBO getCommentReplyList(Integer productId) {
    // 校验商品编号
    Assertion.isPositive(productId, "商品编号不能为空");
    // 根据商品编号查询并按时间升序排列
    Example example = new Example(CommentReply.class);
    example.createCriteria().andEqualTo("productId", productId);
    example.orderBy("createTime");
    // 调用通用查询并将结果集返回给客户端
    return ResponseBO.success(commentRelyService.findList(example));
}
```

（2）CommentReplyServiceImpl 类继承 BaseServiceImpl 类并调用父类 findList 方法查询，同商品类别列表接口查询，重复代码此处不做展示。

（3）出参直接采用数据库对象实体 CommentReply。

8. 返回报文

评论/回复列表接口返回报文如表 21-10 所示。

表 21-10 评论/回复列表接口返回报文说明

参数名	必需	类型	字节长度	说明
id	是	int	11	评论/回复主键编号
productId	是	int	11	商品编号
content	是	string	500	评论/回复内容
fromUserId	是	string	32	回复人用户编号
fromUserName	是	string	50	回复人用户昵称
fromUserAvatar	是	string	500	回复人用户头像
createTime	是	long	13	评论/回复时间
type	是	int	2	类型：1——评论，2——回复
toUserId	是	string	32	被回复人用户编号
toUserName	是	string	50	被回复人用户昵称
toUserAvatar	是	string	500	被回复人用户头像
replyId	是	int	11	回复编号

9. 返回示例

```
{
"code":0,
"message":"success",
"requestId":null,
"timestamp":1567301655000,
"result":[
{
"id":2,
"productId":1,
"content":"测试",
"fromUserId":"xxx",
"fromUserName":"xxx",
"fromUserAvatar":"xxx",
"createTime":1567301655000,
"toUserId":"xxx",
"toUserName":"xxx",
"toUserAvatar":"xxx",
```

```
"replyId":1
}
]
}
```

21.8 点赞 / 取消点赞接口

1. 接口用途

此接口用于用户对商品进行点赞或取消点赞两个功能。

2. 参数校验

校验用户令牌是否为空，校验商品编号是否为空，校验点赞状态是否为空。

3. 业务规则

根据 token 查询用户是否存在，即 token 是否有效；根据商品编号查询商品是否存在；回复目标用户与回复编号要么同时传（回复），要么同时不传（评论）。

4. 请求方式

采用 POST 请求。

5. 请求报文

点赞 / 取消点赞接口请求报文如表 21-11 所示。

表 21-11 点赞 / 取消点赞接口请求报文说明

参数名	必需	类型	字节长度	说明
productId	是	int	11	商品编号
status	是	int	2	点赞状态：1——点赞，2——取消点赞

6. 请求示例

```
{
"productId":1,
"status":1
}
```

7. 编写程序

（1）PraiseController 类，点赞 / 取消点赞相关的接口控制层，praiseOrUnPraise

方法用于接收客户端点赞或取消点赞请求。主要包含校验用户令牌、商品编号、点赞状态是否为空，然后调用 service 层业务，将报文返回给客户端。

```java
/**
 * 点赞/取消点赞
 * @param vo
 * @return
 */
@RequestMapping(value = "praiseOrUnPraise", method = RequestMethod.
POST, produces = MediaType.APPLICATION_JSON_UTF8_VALUE)
public ResponseBO praiseOrUnPraise(@Valid @RequestBody PraiseVO vo) {
    // 校验 token 是否为空
    String token = checkToken();
    vo.setToken(token);
    // 点赞状态枚举值只能为 1 或者 2
    Assertion.isTrue(1 == vo.getStatus() || 2 == vo.getStatus(), "点赞状态异常");
    // 调用 service 层点赞业务
    praiseService.praiseOrUnPraise(vo);
    return ResponseBO.success();
}
```

（2）PraiseServiceImpl 类，praiseOrUnPraise 方法用于点赞/取消点赞业务。校验 token 有效性并返回用户实体，若校验不通过，则返回相应的错误信息。否则继续判断商品信息是否存在，若不存在，则返回相应的错误信息。否则通过用户编号及商品编号查询是否存在点赞记录，若存在点赞记录，则判断传入的点赞状态与数据库记录是否一样。若一样，说明是重复操作，抛出对应的异常信息；若不一样，更新数据库点赞状态。如果不存在点赞记录且请求报文中点赞状态为 2——取消点赞，则说明参数异常，否则插入对应的点赞信息到数据库中，完成点赞/取消点赞业务。

```java
/**
 * 点赞/取消点赞
 * @param vo
 */
@Override
public void praiseOrUnPraise(PraiseVO vo) {
    // 校验 token 是否有效
```

```java
        User user = userService.checkToken(vo.getToken());
        // 查询商品是否存在
        Product product = productMapper.selectByPrimaryKey(vo.getProductId());
        Assertion.notNull(product, "商品不存在");
        // 查询是否存在点赞记录
        Example example = new Example(Praise.class);
        example.createCriteria()
                .andEqualTo("userId", user.getUserId())
                .andEqualTo("productId", vo.getProductId());
        List<Praise> list = praiseMapper.selectByExample(example);
        Date date = new Date();
        if (Detect.notEmpty(list)) {
            Praise praise = Detect.firstOne(list);
            // 已存在点赞记录
            if (null != praise) {
                // 重复点赞或重复取消点赞
                Assertion.notEquals(praise.getStatus(), vo.getStatus(), "请勿重复操作");
                BeanUtils.copyProperties(vo, praise);
                praise.setPraiseTime(date);
                praise.setUserAvatar(user.getUserAvatar());
                praise.setUserName(user.getUserName());
                praiseMapper.updateByPrimaryKeySelective(praise);
                return;
            }
        }
        // 不存在点赞记录，但是转入的值为取消点赞，参数异常
        Assertion.notEquals(vo.getStatus(), 2, "点赞状态异常");
        // 不存在点赞记录，需要新增
        Praise praise = new Praise();
        praise.setProductId(vo.getProductId());
        praise.setUserId(user.getUserId());
        praise.setPraiseTime(date);
        praise.setUserAvatar(user.getUserAvatar());
        praise.setUserName(user.getUserName());
        praiseMapper.insertSelective(praise);
    }
```

（3）入参实体 PraiseVO，对应客户端请求报文。

```java
package com.xpwu.secondary.vo;

import lombok.Getter;
import lombok.Setter;
import lombok.ToString;

import javax.validation.constraints.NotNull;

/**
 * Created by IntelliJ IDEA.
 *
 * @author:caoxue
 * @date:2019/8/15 16:13
 * @description:点赞/取消点赞接口入参
 * @version:1.0
 */
@Setter
@Getter
@ToString(callSuper = true)
public class PraiseVO extends BaseVO {

    private static final long serialVersionUID = 3892995581269313674L;

    /**
     * 商品编号
     */
    @NotNull(message = "商品编号不能为空")
    private Integer productId;

    /**
     * 点赞状态 1——点赞 2——取消点赞
     */
    @NotNull(message = "状态不能为空")
    private Integer status;

}
```

8. 返回报文及示例

参见公共返回示例。

21.9 点赞列表接口

1. 接口用途

此接口用于查询商品相关的点赞列表数据。

2. 参数校验

校验商品编号是否为空。

3. 业务规则

当未登录时,当前商品对当前登录用户的点赞状态默认为未点赞;当用户登录时,查询当前用户是否对该商品点赞;不管是否登录,该商品的所有有效点赞记录全部返回。

4. 请求方式

采用 GET 请求。

5. 请求报文

点赞列表接口请求报文如表 21-12 所示。

表 21-12 点赞列表接口请求报文说明

参数名	必需	类型	字节长度	说明
productId	否	int	11	商品编号

6. 请求示例

GET 请求,将 productId 放在接口 URL 后面即可。

7. 编写程序

(1) PraiseController 类为点赞相关的接口控制层,getPraiseList 方法用于查询点赞列表数据,并校验请求参数商品编号,调用 service 层业务,将报文返回给客户端。

```
/**
 * 获取点赞列表
 * @param vo
 * @return
 */
@RequestMapping(value = "getPraiseList", method = RequestMethod.GET)
public ResponseBO getPraiseList(PraiseVO vo) {
    // 与点赞接口共用入参实体,此处不通过 @Valid 注解与 @NotNull 校验参数,通过断言校验
    Assertion.notNull(vo, "请求参数不能为空");
```

```java
        Assertion.isPositive(vo.getProductId(), "商品编号不能为空");
        String token = getToken();
        vo.setToken(token);
        // 调用 service 层查询
        return ResponseBO.success(praiseService.getPraiseList(vo));
    }
```

（2）PraiseServiceImpl 类，getPraiseList 方法用于查询点赞列表业务。首先根据商品编号查询点赞表有效记录，并按时间升序排列。此接口还需要增加返回一个 praiseStatus 字段，用于客户端判断商品的点赞按钮的亮灭展示。首先判断 token 是否为空，若为空，说明用户未登录，默认为未点赞；若 token 不为空，继续校验其有效性，并获取对应的用户信息。然后根据用户编号及商品编号查询当前登录用户是否对该商品有点赞记录，并返回相应的值给客户端。

```java
    /**
     * 获取点赞列表
     * @param vo
     * @return
     */
    @Override
    public Map<String, Object> getPraiseList(PraiseVO vo) {
        // 根据商品编号查询出有效点赞数据，并按点赞时间升序排列
        Example example = new Example(Praise.class);
        example.createCriteria().andEqualTo("productId", vo.getProductId())
                .andEqualTo("status", 1);
        example.orderBy("praiseTime");
        Map<String, Object> map = new HashMap<>(16);
        map.put("list", praiseMapper.selectByExample(example));
        // 返回值 praiseStatus 字段为当前商品是否被当前用户点赞，客户端用于判断页面爱心是否点亮
        // 若 token 为空，说明未登录，默认为未点赞
        if (!Detect.notEmpty(vo.getToken())) {
            map.put("praiseStatus", 2);
            return map;
        }
        // 若 token 不为空（已登录），校验有效性
        User user = userService.checkToken(vo.getToken());
```

```
    example.clear();
    // 查询当前登录用户是否对该商品有点赞记录
    example.createCriteria().andEqualTo("productId", vo.getProductId())
            .andEqualTo("userId", user.getUserId());
    List<Praise> list = praiseMapper.selectByExample(example);
    // 无点赞记录，点赞状态为未点赞
    if (!Detect.notEmpty(list)) {
        map.put("praiseStatus", 2);
        return map;
    }
    Praise praise = Detect.firstOne(list);
    if (null == praise) {
        map.put("praiseStatus", 2);
        return map;
    }
    // 存在点赞记录，将数据库点赞状态赋值给 praiseStatus 并返回
    map.put("praiseStatus", praise.getStatus());
    return map;
}
```

（3）出参直接采用 Map 对象，见下面的返回报文及示例。

8. 返回报文

点赞列表接口返回报文如表 21-13 所示。

表 21-13　点赞列表接口返回报文说明

参数名	必需	类型	字节长度	说明
praiseStatus	是	int	2	当前点赞状态：1——已点赞，2——未点赞
id	是	int	11	点赞主键编号
productId	是	int	11	商品编号
userId	是	string	32	点赞人用户编号
userName	是	string	50	点赞人用户昵称
userAvatar	是	string	500	点赞人用户头像
status	是	int	2	点赞状态：1——点赞，2——取消点赞
type	是	int	2	类型：1——评论，2——回复
praiseTime	是	long	13	点赞时间

9. 返回示例

```
{
"code":0,
"message":"success",
"requestId":null,
"timestamp":1567301655000,
"result":{
"praiseStatus":1,
"list":[
   {
     "id":2
     "productId":1,
    "userId":"xxx",
     "userName":"xxx",
     "userAvatar":"xxx",
     "status":1,
     "createTime":1567301655000
    }
  ]
 }
}
```

21.10 首页轮播商品列表接口

1. 接口用途

此接口用于查询首页轮播商品列表数据。

2. 参数校验

无参查询。

3. 业务规则

未交易的商品数据，按照想要人数降序排列，取前 5 个商品信息返回。

4. 请求方式

采用 GET 请求。

5. 请求报文

无。

6. 请求示例

无。

7. 编写程序

（1）ProductController 类为商品相关的接口控制层，getBannerList 方法用于查询首页轮播商品列表数据，直接调用 service 层业务，将报文返回给客户端。

```java
/**
 * 获取首页轮播商品列表
 * @return
 */
@RequestMapping(value = "getBannerList", method = RequestMethod.GET)
public ResponseBO getBannerList() {
    return ResponseBO.success(productService.getBannerList());
}
```

（2）ProductServiceImpl 类，getBannerList 方法用于查询首页轮播商品列表业务，直接调用 Mapper 层查询。

```java
/**
 * 获取首页轮播商品列表
 * @return
 */
@Override
public List<ProductBO> getBannerList() {
    // 自定义查询轮播商品信息
    List<ProductBO> list = productMapper.selectBannerList();
    // 通过 filter 将商品图片取第一个返回
    list.stream().filter(o -> {
        String productImgs = o.getProductImgs();
        String [] imgs = productImgs.split(",");
        o.setProductImgs(imgs[0]);
        return true;
    }).collect(Collectors.toList());
    return list;
}
```

（3）ProductMapper 类，自定义 selectBannerList 接口查询。

```
/**
```

```
 * 查询首页轮播商品信息
 * @return
 */
List<ProductBO> selectBannerList();
```

（4）ProductMapper.xml，此处属于扩展查询，需要另外写 SQL 语句连接，具体如下。

```xml
<!-- 首页轮播商品查询 -->
<select id="selectBannerList" resultType="com.xpwu.secondary.bo.ProductBO">
    SELECT
        p.id,
        p.product_imgs productImgs,
        p.product_desc productDesc,
        p.product_price productPrice,
        p.product_type_id productTypeId,
        IFNULL(t.wantNum,0) wantNum,
        pt.'name' productTypeName,
        p.publish_user_id publishUserId,
        u.user_name publishUserName,
        u.user_avatar publishUserAvatar,
        p.create_time createTime,
        p.product_address productAddress
    FROM
        product p
    LEFT JOIN product_type pt ON p.product_type_id = pt.id
    LEFT JOIN 'order' o ON p.id = o.product_id
    LEFT JOIN (SELECT COUNT(1) wantNum,product_id FROM chat c GROUP BY product_id) t ON p.id = t.product_id
    LEFT JOIN 'user' u ON p.publish_user_id = u.user_id
    WHERE (o.order_id is null or o.trade_status = 2)
    GROUP BY p.id
    ORDER BY t.wantNum DESC LIMIT 5
</select>
```

（5）返回报文实体 ProductBO。

```
package com.xpwu.secondary.bo;
```

```java
import lombok.Getter;
import lombok.Setter;
import lombok.ToString;

import java.io.Serializable;
import java.math.BigDecimal;
import java.util.Date;

/**
 * Created by IntelliJ IDEA.
 *
 * @author:caoxue
 * @date:2019/9/12 9:55
 * @description:
 * @version:1.0
 */
@Setter
@Getter
@ToString(callSuper = true)
public class ProductBO implements Serializable {

    private static final long serialVersionUID = 8821081481707766726L;

    /**
     * 商品编号
     */
    private Integer id;

    /**
     * 商品图片,多个图片地址用英文逗号隔开
     */
    private String productImgs;

    /**
     * 商品描述
     */
    private String productDesc;

    /**
     * 商品价格
     */
```

```java
    private BigDecimal productPrice;

    /**
     * 商品类别编号
     */
    private Integer productTypeId;

    /**
     * 商品类别名称
     */
    private String productTypeName;

    /**
     * 商品发布人用户编号
     */
    private String publishUserId;

    /**
     * 商品发布人用户昵称
     */
    private String publishUserName;

    /**
     * 商品发布人用户头像
     */
    private String publishUserAvatar;

    /**
     * 购买用户编号
     */
    private String buyingUserId;

    /**
     * 交易状态：1——未交易，2——交易中，3——交易成功，4——交易失败
     */
    private Integer tradeStatus;

    /**
     * 创建时间（发布时间）
     */
    private Date createTime;
```

```java
/**
 * 交易时间(购买时间)
 */
private Date tradeTime;

/**
 * 想要人数
 */
private Integer wantNum;

/**
 * 订单编号
 */
private Integer orderId;

/**
 * 商品地址
 */
private String productAddress;

}
```

8. 返回报文

首页轮播商品列表接口报文如表 21-14 所示。

表 21-14　首页轮播商品列表接口返回报文说明

参数名	必需	类型	字节长度	说明
id	是	int	11	商品编号
productImgs	是	string	2000	商品图片，多个图片地址用英文逗号隔开
productDesc	是	string	500	商品描述
productPrice	是	double	10	商品价格
productTypeId	是	Int	11	商品类别编号
productTypeName	是	string	100	商品类别名称
publishUserId	是	string	32	商品发布人用户编号
publishUserName	是	string	50	商品发布人用户昵称
publishUserAvatar	是	string	100	商品发布人用户头像
createTime	是	long	13	商品发布时间
wantNum	是	int	11	想要人数（有多少人想要此商品）
productAddress	是	string	500	商品地址

9. 返回示例

```
{
"code":0,
"message":"success",
"requestId":null,
"timestamp":1567301655000,
"result":[
{
"id":1,
"productImgs":"xxx",
"productDesc":"全新手机",
"productPrice":4000,
"productTypeId":1,
"productTypeName":"手机",
"publishUserId":"xxx",
"publishUserName":"xxx",
"publishUserAvatar":"xxx",
"createTime":1567301655000,
"wantNum":5,
"productAddress":"xxx"
}
]
}
```

第 22 章 消息模块接口

【本章导读】

◎ 聊天对话框列表接口

◎ 获取聊天详情接口

◎ 初始化聊天接口

◎ 发送消息接口

22.1 聊天对话框列表接口

1. 接口用途

此接口用于查询用户消息页面的聊天对话框列表数据。

2. 参数校验

校验用户令牌是否为空。

3. 业务规则

校验 token 是否有效，若无效，抛出对应异常信息；若有效，开启分页查询，根据用户编号查询该用户所有的聊天对话框列表数据。

4. 请求方式

采用 POST 请求。

5. 请求报文

聊天对话框列表接口请求报文如表 22-1 所示。

表 22-1 聊天对话框列表接口请求报文说明

参数名	必需	类型	字节长度	说明
pageNum	否	int	11	当前页码，不传默认为 1
pageSize	否	int	11	每页条数，不传默认为 10

6. 请求示例

```
{
"pageNum":1,
"pageSize":10
}
```

7. 编写程序

（1）ChatController 类为聊天相关接口的控制层，getChatList 方法用于查询聊天对话框列表数据。主要包含校验 token 是否为空，并调用 service 层业务，将报文返回给客户端。

```
/**
 * 查询聊天对话框列表数据
 * @param vo
 * @return
 */
@RequestMapping(value = "getChatList", method = RequestMethod.POST)
public ResponseBO getChatList(BaseVO vo) {
    // 校验token是否为空
    String token = checkToken();
    vo.setToken(token);
    // 调用service层业务
    return ResponseBO.successPageInfo(chatService.getChatList(vo));
}
```

（2）ChatServiceImpl 类，getChatList 方法用于查询聊天对话框列表业务。首先校验 token 是否有效，若有效，获取对应的用户信息，然后根据用户编号查询聊天对话框列表数据，并开启分页查询，最后返回相应的结果集给客户端。

```
/**
 * 查询聊天对话框列表数据
 *
 * @param vo
 * @return
 */
```

```java
@Override
public List<ChatBO> getChatList(BaseVO vo) {
    // 校验token
    User user = userService.checkToken(vo.getToken());
    // 开启分页查询
    PageHelper.startPage(vo.getPageNum(), vo.getPageSize());
    // 调用mapper层查询数据
    return chatListMapper.selectChatList(user.getUserId());
}
```

（3）ChatListMapper.xml，此查询为扩展查询，连接多表得到结果集，具体 SQL 语句如下。

```xml
<!-- 查询聊天对话框列表数据 -->
<select id="selectChatList" parameterType="string" resultType="com.xpwu.secondary.bo.ChatBO">
    SELECT
        cl.chat_id chatId,
        cl.user_id userId,
        cl.user_name userName,
        cl.user_avatar userAvatar,
        cl.another_user_id anotherUserId,
        cl.another_user_name anotherUserName,
        cl.another_user_avatar anotherUserAvatar,
        cl.unread,
        cl.create_time createTime,
        cl.update_time updateTime,
        cd.content lastChatContent,
        p.id productId,
        p.product_imgs productImgs,
        p.product_price productPrice
    FROM chat_list cl
    LEFT JOIN chat_detail cd ON cl.chat_id = cd.chat_id AND cd.is_latest = 1
    LEFT JOIN product p ON p.id = cl.product_id
    WHERE cl.'status' = 1 AND cl.user_id = #{userId}
    ORDER BY cl.update_time DESC
</select>
```

（4）出参实体 ChatBO，此处在继承 ChatList 实体的基础上增加了如下属性。

```
package com.xpwu.secondary.bo;

import com.xpwu.secondary.entity.ChatList;
import lombok.Getter;
import lombok.Setter;
import lombok.ToString;

import java.io.Serializable;
import java.math.BigDecimal;

/**
 * Created by IntelliJ IDEA.
 *
 * @author:caoxue
 * @date:2019/8/20 21:06
 * @description:
 * @version:1.0
 */
@Setter
@Getter
@ToString(callSuper = true)
public class ChatBO extends ChatList implements Serializable {

    private static final long serialVersionUID = 8379355470056464837L;

    /**
     * 最后一条消息
     */
    private String lastChatContent;

    /**
     * 商品编号
     */
    private Integer productId;

    /**
     * 商品图片
     */
```

```
    private String productImgs;

    /**
     * 商品价格
     */
    private BigDecimal productPrice;

}
```

8. 返回报文

聊天对话框列表接口返回报文如表 22-2 所示。

表 22-2　聊天对话框列表接口返回报文说明

参数名	必需	类型	字节长度	说明
chatId	是	int	11	聊天主表编号
userId	是	string	32	用户编号
userName	是	string	50	点赞人用户昵称
userAvatar	是	string	500	点赞人用户头像
anotherUserId	是	string	32	对方用户编号
anotherUserName	是	string	50	对方用户昵称
anotherUserAvatar	是	string	500	对方用户头像
lastChatContent	是	string	500	最后一条消息
unread	是	int	4	未读数
createTime	是	long	13	创建时间
updateTime	是	long	13	最后一条消息发送时间
productId	是	int	11	商品编号
productImgs	是	string	2000	商品图片
productPrice	是	double	10	商品价格

9. 返回示例

```
{
"code":0,
"message":"success",
"requestId":null,
"timestamp":1567301655000,
```

```
"result":[
   {
     "chatId":1
    "userId":"xxx",
     "userName":"xxx",
     "userAvatar":"xxx",
    "anotherUserId":"xxx",
     "anotherUserName":"xxx",
     "anotherUserAvatar":"xxx",
    "lastChatContent":"xxx",
     "unread":1,
     "createTime":1567301655000,
     "updateTime":1567301655000,
     "productId":1,
     "productImgs":"xxx",
     "productPrice":100
   }
  ]
}
```

22.2 获取聊天详情接口

1. 接口用途

此接口用于查询用户与用户聊天详情页的消息记录，支持分页查询。

2. 参数校验

校验用户令牌是否为空，校验聊天主表编号是否为空。

3. 业务规则

校验 token 是否有效，若无效，抛出对应的异常信息；若有效，则将该聊天记录的未读数更新为 0（用户打开聊天记录时，未读数更新为 0）。根据聊天主表编号查询该聊天的聊天记录数据，并按发送时间降序排列。

4. 请求方式

采用 POST 请求。

5. 请求报文

获取聊天详情接口请求报文如表 22-3 所示。

表 22-3 获取聊天详情接口请求报文说明

参数名	必需	类型	字节长度	说明
pageNum	否	int	11	当前页码，不传默认为 1
pageSize	否	int	11	每页条数，不传默认为 10
chatId	是	string	32	聊天主表编号

6. 请求示例

```
{
"pageNum":1,
"pageSize":10,
"chatId":"xxx"
}
```

7. 编写程序

（1）ChatController 类为聊天相关的接口控制层，getChatDetailList 方法用于查询聊天详情列表数据。主要包含校验 token 是否为空，并调用 service 层业务，将报文返回给客户端。

```
    /**
     * 查询聊天详情
     * @param vo
     * @return
     */
    @RequestMapping(value = "getChatDetailList", method = RequestMethod.POST)
    public ResponseBO getChatDetailList(@RequestBody ChatDetailVO vo) {
        // 校验 token 是否为空
        String token = checkToken();
        vo.setToken(token);
        // 校验聊天主表编号是否为空
        Assertion.notEmpty(vo.getChatId(), "chatId 不能为空 ");
        // 调用 service 层查询
        return ResponseBO.successPageInfo(chatService.getChatDetailList(vo));
    }
```

（2）ChatServiceImpl 类，getChatDetailList 方法用于查询聊天详情列表数据业务。首先校验 token 是否有效，若有效，获取对应的用户信息，根据用户编号及聊天主表编号将未读数更新为 0。然后开启分页查询，根据聊天主表编号查询聊天详情表，并按发送时间降序排列。最后返回聊天记录数据给客户端。

```java
/**
 * 查询聊天详情
 *
 * @param vo
 * @return
 */
@Override
public List<ChatDetail> getChatDetailList(ChatDetailVO vo) {
    // 校验token
    User user = userService.checkToken(vo.getToken());
    ChatList chatList = new ChatList();
    // 将未读数变为0
    chatList.setUnread(0);
    Example chatListExample = new Example(ChatList.class);
    chatListExample.createCriteria().andEqualTo("chatId", vo.getChatId())
            .andEqualTo("anotherUserId", user.getUserId());
    // 更新未读数量
    chatListMapper.updateByExampleSelective(chatList, chatListExample);
    // 开启分页查询聊天详情数据
    PageHelper.startPage(vo.getPageNum(), vo.getPageSize());
    Example example = new Example(ChatDetail.class);
    example.createCriteria().andEqualTo("chatId", vo.getChatId());
    // 降序排列
    example.setOrderByClause(" create_time desc");
    // 调用通用查询
    return chatDetailMapper.selectByExample(example);
}
```

（3）入参实体 ChatDetailVO，主要对应请求报文。

```java
package com.xpwu.secondary.vo;

import lombok.Getter;
import lombok.Setter;
```

```
import lombok.ToString;

/**
 * Created by IntelliJ IDEA.
 *
 * @author:caoxue
 * @date:2019/8/24 23:35
 * @description:查询聊天详情入参
 * @version:1.0
 */
@Setter
@Getter
@ToString(callSuper = true)
public class ChatDetailVO extends BaseVO {

    private static final long serialVersionUID = 8732608165375370704L;

    /**
     * 聊天主表编号
     */
    private String chatId;

}
```

（4）出参为 ChatDetail 数据库实体，通过 MyBatis 生成工具生成。

```
package com.xpwu.secondary.entity;

import com.fasterxml.jackson.annotation.JsonIgnore;
import lombok.Getter;
import lombok.Setter;

import javax.persistence.*;
import java.util.Date;

@Setter
@Getter
@Table(name = "'chat_detail'")
public class ChatDetail {
    /**
     * 消息详情
```

```java
     */
    @Id
    @Column(name = "'id'")
    @GeneratedValue(strategy = GenerationType.IDENTITY)
    @JsonIgnore
    private Integer id;

    /**
     * 聊天主表编号
     */
    @Column(name = "'chat_id'")
    @JsonIgnore
    private String chatId;

    /**
     * 消息发送者用户编号
     */
    @Column(name = "'user_id'")
    private String userId;

    /**
     * 消息发送者用户昵称
     */
    @Column(name = "'user_name'")
    private String userName;

    /**
     * 消息发送者用户头像
     */
    @Column(name = "'user_avatar'")
    private String userAvatar;

    /**
     * 消息内容
     */
    @Column(name = "'content'")
    private String content;

    /**
     * 消息类型：1——用户消息；2——系统消息
     */
```

```java
    @Column(name = "'type'")
    @JsonIgnore
    private Integer type;

    /**
     * 是否是最后一条消息：1——是；2——否
     */
    @Column(name = "'is_latest'")
    private Integer isLatest;

    /**
     * 创建时间（发送时间）
     */
    @Column(name = "'create_time'")
    private Date createTime;

    /**
     * 修改时间
     */
    @Column(name = "'update_time'")
    @JsonIgnore
    private Date updateTime;

}
```

8. 返回报文

获取聊天详情接口返回报文如表 22-4 所示。

表 22-4　获取聊天详情接口返回报文说明

参数名	必需	类型	字节长度	说明
userId	是	string	32	消息发送者用户编号
userName	是	string	50	消息发送者用户昵称
userAvatar	是	string	200	消息发送者用户头像
content	是	string	500	消息内容
isLatest	是	int	2	是否是最后一条消息：1——是，2——否
createTime	是	long	13	发送时间

9. 返回示例

{

```
"code":0,
"message":"success",
"requestId":null,
"timestamp":1567301655000,
"result":[
   {
    "userId":"xxx",
     "userName":"xxx",
     "userAvatar":"xxx",
    "content":"xxx",
     "isLatest":2,
     "createTime":1567301655000
   }
  ]
}
```

22.3 初始化聊天接口

1. 接口用途

此接口用于用户在查看商品点击【我想要】时发起请求，服务端初始化聊天数据。

2. 参数校验

校验用户令牌是否为空，校验发送对象用户编号是否为空，校验商品编号是否为空。

3. 业务规则

校验 token 是否有效，查询商品信息是否存在，查询聊天主表是否有记录，初始化聊天数据（聊天主表插入一行记录，聊天对话框列表插入两行记录）。

4. 请求方式

采用 POST 请求。

5. 请求报文

初始化聊天接口请求报文如表 22-5 所示。

表 22-5　初始化聊天接口请求报文说明

参数名	必需	类型	字节长度	说明
toUserId	是	string	32	发送对象用户编号
productId	是	int	11	商品编号

6. 请求示例

```
{
"toUserId":"xxx",
"productId":1
}
```

7. 编写程序

(1) ChatController 类为聊天相关的接口控制层,initChat 方法用于接收客户端点击【我想要】请求,初始化聊天数据。主要包含校验用户令牌、商品编号、发送对象用户编号是否为空,并调用 service 层业务,将报文返回给客户端。

```java
/**
 * 初始化聊天
 * @param vo
 * @return
 */
@RequestMapping(value = "initChat", method = RequestMethod.POST)
public ResponseBO initChat(@RequestBody InitChatVO vo) {
    // 校验token是否为空
    String token = checkToken();
    vo.setToken(token);
    // 校验发送对象用户编号是否为空
    Assertion.notEmpty(vo.getToUserId(), "toUserId 不能为空 ");
    // 校验商品编号是否为空
    Assertion.isPositive(vo.getProductId(), " 商品编号不能为空 ");
    // 调用 service 层业务得到聊天主表编号
    String chatId = chatService.init(vo);
    Map<String, String> map = new HashMap<>(16);
    map.put("chatId", chatId);
    return ResponseBO.success(map);
}
```

(2) ChatServiceImpl 类,init 方法用于初始化聊天业务。首先校验 token 是否有效,若有效,获取对应的用户信息。然后查询商品信息是否存在,根据商品编号、当前登录人用户编号、发送对象用户编号等参数查询聊天记录是否存在。若不存在,初始化聊天,向聊天主表中插入一行记录,同时向聊天列表中插入两行记录。其分别表示聊天者与被聊天者聊天对话框列表的数据,即 userId 与 toUserId 对调,一行代表 A 用户聊天界面的聊天对话框,另一行代表 B 用户界面的聊天对话框。若聊天主表存在记录,则直接返回聊天主表编号。

```java
/**
 * 初始化聊天
 *
 * @param vo
 * @return
 */
@Override
public String init(InitChatVO vo) {
    // 校验 token 是否有效
    User user = userService.checkToken(vo.getToken());
    // 查询商品信息
    Product product = productMapper.selectByPrimaryKey(vo.getProductId());

    Assertion.notNull(product, "商品信息不存在");
    // 查询聊天主表是否有记录
    Example example = new Example(Chat.class);
    example.createCriteria().andEqualTo("userId", user.getUserId())
            .andEqualTo("anotherUserId", vo.getToUserId())
            .andEqualTo("productId", vo.getProductId());
    List<Chat> list = chatMapper.selectByExample(example);
    String chatId = null;
    if (!Detect.notEmpty(list)) {
        // 将用户编号与对方用户编号对调查询(双方点击都有可能产生聊天记录,所以需要对调参数查询),再次确认是否有聊天记录
        Example example2 = new Example(Chat.class);
        example2.createCriteria().andEqualTo("anotherUserId", user.getUserId())
                .andEqualTo("userId", vo.getToUserId())
                .andEqualTo("productId", vo.getProductId());
        list = chatMapper.selectByExample(example2);
        if (!Detect.notEmpty(list)) {
            // 若聊天主表无记录,属于第一次点击,则初始化聊天
            User toUser = userService.findById(vo.getToUserId());
            Assertion.notNull(toUser, "对方用户不存在");
            chatId = UUIDUtils.getUid();
```

```java
                    vo.setToUserName(toUser.getUserName());
                    vo.setToUserAvatar(toUser.getUserAvatar());
                    initChat(vo, chatId, vo.getProductId(), user, new Date(), 0);
                    // 返回聊天编号给客户端
                    return chatId;
                }
            }
            // 已存在聊天记录，直接返回聊天编号
            Chat chat = Detect.firstOne(list);
            if (chat != null) {
                chatId = chat.getId();
            }
            return chatId;
        }

        /**
         * 初始化聊天数据
         *
         * @param vo
         * @param chatId
         * @param productId
         * @param user
         * @param date
         * @param unread
         */
        private void initChat(ChatSocketVO vo, String chatId, Integer productId, User user, Date date, Integer unread) {
            // 向聊天主表插入记录
            Chat chat = new Chat();
            chat.setAnotherUserId(vo.getToUserId());
            chat.setUserId(user.getUserId());
            chat.setId(chatId);
            chat.setProductId(productId);
            chatMapper.insertSelective(chat);
            // 聊天列表新增2条记录,1条为A用户看到的记录、1条为B用户看到的记录
            ChatList chatList = new ChatList();
            chatList.setAnotherUserId(vo.getToUserId());
            chatList.setAnotherUserAvatar(vo.getToUserAvatar());
            chatList.setAnotherUserName(vo.getToUserName());
            chatList.setCreateTime(date);
```

```java
chatList.setUpdateTime(date);
chatList.setStatus(1);
// 在线
if (chatSocket.online(user.getUserId())) {
    chatList.setIsOnline(1);
} else {
    // 不在线
    chatList.setIsOnline(2);
}
chatList.setUserAvatar(user.getUserAvatar());
chatList.setUserId(user.getUserId());
chatList.setUserName(user.getUserName());
chatList.setChatId(chatId);
if (Detect.isPositive(productId)) {
    chatList.setProductId(productId);
}
// A用户聊天列表记录
chatListMapper.insertSelective(chatList);
ChatList secondChatList = new ChatList();
secondChatList.setUserId(vo.getToUserId());
secondChatList.setUserAvatar(vo.getToUserAvatar());
secondChatList.setUserName(vo.getToUserName());
secondChatList.setAnotherUserId(user.getUserId());
secondChatList.setAnotherUserAvatar(user.getUserAvatar());
secondChatList.setAnotherUserName(user.getUserName());
// 不在线
if (!chatSocket.online(vo.getToUserId())) {
    secondChatList.setUnread(unread);
    secondChatList.setIsOnline(2);
} else {
    // 在线
    secondChatList.setIsOnline(1);
}
secondChatList.setChatId(chatId);
secondChatList.setCreateTime(date);
secondChatList.setUpdateTime(date);
secondChatList.setStatus(1);
if (Detect.isPositive(productId)) {
    secondChatList.setProductId(productId);
}
// B用户聊天列表记录
```

```
        chatListMapper.insertSelective(secondChatList);
    }
```

（3）入参实体 InitChatVO，主要对应请求报文。考虑到发送消息接口也要调用上述 initChat 方法，所以这里用 InitChatVO 继承 ChatSocketVO（发送消息接口入参），这样两边就可以通过各自 VO 对象调用私有方法 initChat（ChatSocketVO vo,...）方法。

```
package com.xpwu.secondary.vo;

import lombok.Getter;
import lombok.Setter;
import lombok.ToString;

/**
 * Created by IntelliJ IDEA.
 *
 * @author:caoxue
 * @date:2019/9/13 20:09
 * @description:初始化聊天（点击【我想要】）入参
 * @version:1.0
 */
@Setter
@Getter
@ToString(callSuper = true)
public class InitChatVO extends ChatSocketVO {

    private static final long serialVersionUID = -7080692308595415186L;

    /**
     * 商品编号
     */
    private Integer productId;

}
```

（4）ChatSocketVO 实体为发送聊天消息接口入参。

```
package com.xpwu.secondary.vo;
```

```java
import lombok.Getter;
import lombok.Setter;
import lombok.ToString;

/**
 * Created by IntelliJ IDEA.
 *
 * @author:caoxue
 * @date:2019/8/18 23:49
 * @description: 发送聊天入参
 * @version:1.0
 */
@Setter
@Getter
@ToString(callSuper = true)
public class ChatSocketVO extends BaseVO {

    private static final long serialVersionUID = 5720259669569605483L;

    /**
     * 发送对象用户编号
     */
    private String toUserId;

    /**
     * 发送对象用户昵称
     */
    private String toUserName;

    /**
     * 发送对象用户头像
     */
    private String toUserAvatar;

    /**
     * 消息内容
     */
    private String content;

    /**
     * 聊天主表编号
```

```
        */
        private String chatId;

}
```

8. 返回报文

初始化聊天接口返回报文如表 22-6 所示。

表 22-6　初始化聊天接口返回报文说明

参数名	必需	类型	字节长度	说明
chatId	是	string	32	聊天主表编号

9. 返回示例

```
{
"code":0,
"message":"success",
"requestId":null,
"timestamp":1567301655000,
"result":
    {
     "chatId":"xxx"
    }
}
```

22.4　发送消息接口

1. 接口用途

此接口用于用户在查看商品时点击【我想要】后实时与对方聊天，发送聊天消息。

2. 参数校验

校验用户令牌等请求报文是否为空。

3. 业务规则

校验 token 是否有效；查询是否为第一次聊天；查询聊天列表数据是否正常；判断对方是否在线，若不在线，对方聊天列表对应数据未读数加 1。

4. 请求方式

采用 WebSocket 调用。

5. 请求报文

发送消息接口请求报文如表 22-7 所示。

表 22-7 发送消息接口请求报文说明

参数名	必需	类型	字节长度	说明
toUserId	是	string	32	对方用户编号
toUserName	是	string	50	对方用户昵称
toUserAvatar	是	string	200	对方用户头像
token	是	string	32	用户令牌
content	是	string	500	消息内容
chatId	是	string	32	聊天主表编号

6. 请求示例

```
{
"toUserId":"xxx",
"toUserName":"xxx",
"toUserAvatar":"xxx",
"token":"xxx",
"content":"xxx",
"chatId":"xxx"
}
```

7. 编写程序

（1）WebSocket 是一种在单个 TCP 连接上进行全双工通信的协议，即长连接。它使得客户端与服务端之间的数据交换变得非常简单，允许服务端主动向客户端推送数据。在 WebSocket API 中，浏览器与服务端只需要完成一次握手，两者之间就可以创建持久性的连接，并进行双向数据传输。本项目正是基于这一特点，采用 WebSocket 实现简单的实时聊天功能。

ChatSocket 类中包含了建立连接、关闭连接、监听客户端消息、错误/异常监听、发送消息等方法。

onOpen：与客户端建立连接。

onClose：关闭当前连接。

onMessage：监听客户端推送来的消息，并且调用 service 层业务将消息入库。

onError：错误/异常监听。

sendMessage：发送消息（基础方法）。

sendInfo：在 sendMessage 基础上封装，发送消息给指定通道（用户）。

online：判断用户是否在线。

getOnlineCount：获取当前在线人数。

addOnlineCount：在线人数加 1，建立连接时调用。

subOnlineCount：在线人数减 1，关闭连接时调用。

具体代码如下。

```java
package com.xpwu.secondary.socket;

import com.alibaba.fastjson.JSON;
import com.xpwu.secondary.vo.ChatSocketVO;
import com.xpwu.secondary.service.ChatService;
import com.xpwu.secondary.utils.JacksonUtils;
import com.xpwu.secondary.utils.SpringContextUtil;
import lombok.EqualsAndHashCode;
import lombok.extern.slf4j.Slf4j;
import org.springframework.stereotype.Component;

import javax.websocket.*;
import javax.websocket.server.ServerEndpoint;
import java.io.IOException;
import java.util.concurrent.ConcurrentHashMap;
import java.util.concurrent.atomic.AtomicInteger;

/**
 * @author caoxue.
 * @date 2019/8/18
 */
@ServerEndpoint("/socket")
@Component
@Slf4j
@EqualsAndHashCode
public class ChatSocket {
```

```java
    /**
     * 静态变量,用来记录当前在线连接数。应该把它设计成线程安全的
     */
    private static AtomicInteger onlineCount = new AtomicInteger(0);

    /**
     * ConcurrentHashMap 是线程安全 k-v 组合集合
     */
    private static ConcurrentHashMap<String, ChatSocket> concurrentHashMap = new ConcurrentHashMap<>(16);

    /**
     * 与某个客户端的连接会话,需要通过它来给客户端发送消息
     */
    private Session session;

    /**
     * 连接建立成功后调用的方法
     */
    @OnOpen
    public void onOpen(Session session) {
        this.session = session;
        // 利用用户编号绑定唯一对应通道
        concurrentHashMap.putIfAbsent(session.getQueryString(), this);
        // 在线人数加 1
        addOnlineCount();
        log.info("有新连接加入【{}】! 当前在线人数为 " + getOnlineCount(), session.getQueryString());
        try {
            sendMessage("有新连接加入! " + session.getId());
        } catch (IOException e) {
            log.error("IO 异常", e);
        }
    }

    /**
     * 连接关闭后调用的方法
     */
    @OnClose
    public void onClose() {
        /* 移除此通道 */
```

```java
        concurrentHashMap.remove(this.session.getQueryString());
        /* 在线人数减 1*/
        subOnlineCount();
        log.info("通道为【" + this.session.getQueryString() + "】的连接关闭！当前在线人数为 " + getOnlineCount());
    }

    /**
     * 收到客户端消息后调用的方法
     *
     * @param message 客户端发送过来的消息
     */
    @OnMessage
    public void onMessage(String message) throws IOException {
        log.info("来自客户端的消息：" + message);
        // 防止将心跳数据入库并发送
        String reg = "{";
        // 消息入库是否成功
        boolean send = false;
        if (message.contains(reg)) {
            // 请求参数
            ChatSocketVO vo = JSON.parseObject(message, ChatSocketVO.class);
            // 消息记录入库
            try {
                // 通过 spring 获取 bean 对象
                ChatService chatService = SpringContextUtil.getBean(ChatService.class);
                // 调用 service 层业务将消息入库
                send = chatService.chat(vo);
            } catch (Exception e) {
                log.warn("消息入库失败 ", e);
            }
            // 入库成功，推送消息到客户端
            if (send) {
                // 发送消息
                sendInfo(vo);
            }
        }
    }
```

```java
/**
 * 发生错误时调用
 *
 * @param error
 */
@OnError
public void onError(Throwable error) {
    log.info("发生错误");
    error.printStackTrace();
}


/**
 * 发送消息——基础服务
 *
 * @param message
 * @throws IOException
 */
private void sendMessage(String message) throws IOException {
    this.session.getBasicRemote().sendText(message);
}



/**
 * 向客户端推送消息
 * @param vo
 * @throws IOException
 */
private static void sendInfo(ChatSocketVO vo) throws IOException {
    // 获取唯一通道
    ChatSocket socket = concurrentHashMap.get(vo.getToUserId());
    String message = JacksonUtils.objectToJson(vo);
    if (socket != null) {
        log.info("开始发送socket消息, message=" + message);
        socket.sendMessage(message);
    } else {
        log.info("未找到socket通道,不发送消息, message=" + message);
    }
}

/**
 * 判断用户是否在线
```

```java
 * @param userId
 * @return
 */
public boolean online(String userId) {
    return concurrentHashMap.containsKey(userId);
}

/**
 * 获取在线人数
 * @return
 */
private static synchronized int getOnlineCount() {
    return onlineCount.get();
}

/**
 * 在线人数加 1
 */
private static synchronized void addOnlineCount() {
    ChatSocket.onlineCount.incrementAndGet();
}

/**
 * 在线人数减 1
 */
private static synchronized void subOnlineCount() {
    ChatSocket.onlineCount.decrementAndGet();
}
```

（2）ChatServiceImpl 类，chat 方法用于发送聊天消息业务。首先校验 token 是否有效，若有效，获取对应的用户信息。然后根据聊天主表编号查询是否有相应的聊天记录，若没有，说明是第一次聊天，调用初始化聊天方法；若不是第一次聊天，判断对方是否在线，若不在线，将未读数加 1。最后将聊天记录插入表中，完成消息入库业务。

```java
/**
 * 发送聊天消息并入库
 * @param vo
 */
```

```java
    @Override
    @Transactional(rollbackFor = Exception.class, isolation = Isolation.READ_COMMITTED)
    public boolean chat(ChatSocketVO vo) {
        // 校验token
        User user = userService.checkToken(vo.getToken());
        // 查询聊天
        Example example = new Example(Chat.class);
        example.createCriteria().andEqualTo("id", vo.getChatId());
        List<Chat> list = chatMapper.selectByExample(example);
        Date date = new Date();
        String chatId;
        // 第一次聊天
        if (!Detect.notEmpty(list)) {
            // 初始化聊天
            chatId = UUIDUtils.getUid();
            initChat(vo, chatId, null, user, date, 1);
        } else {
            // 不是第一次聊天
            Chat chat = Detect.firstOne(list);
            Assertion.notNull(chat, "数据异常,发送失败");
            chatId = chat.getId();
            // 不在线
            if (!chatSocket.online(vo.getToUserId())) {
                // 查询聊天对话框记录,校验数据是否正常
                Example chatListExample = new Example(ChatList.class);
                chatListExample.createCriteria().andEqualTo("chatId", chatId)
                        .andEqualTo("userId", vo.getToUserId());
                List<ChatList> chatLists = chatListMapper.selectByExample(chatListExample);
                Assertion.notEmpty(chatLists, "数据异常,发送失败");
                ChatList chatList = Detect.firstOne(chatLists);
                Assertion.notNull(chatList, "数据异常,发送失败");
                ChatList update = new ChatList();
                update.setUnread(chatList.getUnread() + 1);
                update.setId(chatList.getId());
                update.setUpdateTime(date);
                // 聊天未读数加1并修改入库
                chatListMapper.updateByPrimaryKeySelective(update);
            }
```

```java
        // 将最后一条消息标识去掉
        Example detailExample = new Example(ChatDetail.class);
        detailExample.createCriteria().andEqualTo("chatId", chatId).andEqualTo("isLatest", 1);
        ChatDetail update = new ChatDetail();
        update.setIsLatest(0);
        update.setUpdateTime(date);
        chatDetailMapper.updateByExampleSelective(update, detailExample);
    }
    // 向处理聊天详情记录
    ChatDetail chatDetail = new ChatDetail();
    chatDetail.setChatId(chatId);
    chatDetail.setContent(vo.getContent());
    chatDetail.setCreateTime(date);
    chatDetail.setIsLatest(1);
    chatDetail.setUpdateTime(date);
    chatDetail.setUserAvatar(user.getUserAvatar());
    chatDetail.setUserName(user.getUserName());
    chatDetail.setUserId(user.getUserId());
    // 插入聊天记录
    chatDetailMapper.insertSelective(chatDetail);
    return true;
}

/**
 * 初始化聊天数据
 *
 * @param bo
 * @param chatId
 * @param productId
 * @param user
 * @param date
 * @param unread
 */
private void initChat(ChatSocketBO bo, String chatId, Integer productId, User user, Date date, Integer unread) {
    // 向聊天主表插入记录
    Chat chat = new Chat();
    chat.setAnotherUserId(bo.getToUserId());
    chat.setUserId(user.getUserId());
```

```java
chat.setId(chatId);
chat.setProductId(productId);
chatMapper.insertSelective(chat);
// 聊天列表新增2条记录,1条为A用户看到的记录,1条为B用户看到的记录
ChatList chatList = new ChatList();
chatList.setAnotherUserId(bo.getToUserId());
chatList.setAnotherUserAvatar(bo.getToUserAvatar());
chatList.setAnotherUserName(bo.getToUserName());
chatList.setCreateTime(date);
chatList.setUpdateTime(date);
chatList.setStatus(1);
// 在线
if (chatSocket.online(user.getUserId())) {
    chatList.setIsOnline(1);
} else {
    // 不在线
    chatList.setIsOnline(2);
}
chatList.setUserAvatar(user.getUserAvatar());
chatList.setUserId(user.getUserId());
chatList.setUserName(user.getUserName());
chatList.setChatId(chatId);
if (Detect.isPositive(productId)) {
    chatList.setProductId(productId);
}
// A用户聊天列表记录
chatListMapper.insertSelective(chatList);
ChatList secondChatList = new ChatList();
secondChatList.setUserId(bo.getToUserId());
secondChatList.setUserAvatar(bo.getToUserAvatar());
secondChatList.setUserName(bo.getToUserName());
secondChatList.setAnotherUserId(user.getUserId());
secondChatList.setAnotherUserAvatar(user.getUserAvatar());
secondChatList.setAnotherUserName(user.getUserName());
// 不在线
if (!chatSocket.online(bo.getToUserId())) {
    secondChatList.setUnread(unread);
    secondChatList.setIsOnline(2);
} else {
    // 在线
    secondChatList.setIsOnline(1);
```

```
        }
        secondChatList.setChatId(chatId);
        secondChatList.setCreateTime(date);
        secondChatList.setUpdateTime(date);
        secondChatList.setStatus(1);
        if (Detect.isPositive(productId)) {
            secondChatList.setProductId(productId);
        }
        // B用户聊天列表记录
        chatListMapper.insertSelective(secondChatList);
}
```

（3）ChatSocketVO 实体为发送聊天消息接口入参。

```
package com.xpwu.secondary.vo;

import lombok.Getter;
import lombok.Setter;
import lombok.ToString;

/**
 * Created by IntelliJ IDEA.
 *
 * @author:caoxue
 * @date:2019/8/18 23:49
 * @description:发送聊天入参
 * @version:1.0
 */
@Setter
@Getter
@ToString(callSuper = true)
public class ChatSocketVO extends BaseVO {

    private static final long serialVersionUID = 5720259669569605483L;

    /**
     * 发送对象用户编号
     */
    private String toUserId;

    /**
```

```
 * 发送对象用户昵称
 */
private String toUserName;

/**
 * 发送对象用户头像
 */
private String toUserAvatar;

/**
 * 消息内容
 */
private String content;

/**
 * 聊天主表编号
 */
private String chatId;
}
```

8. 返回报文及示例

无。

第 23 章 支付模块接口

【本章导读】

◎ 购买商品接口

◎ 获取订单详情接口

◎ 取消订单接口

◎ 支付宝 WAP 支付流程接口

◎ 支付宝支付接口

◎ 支付宝支付回调接口

23.1 购买商品接口

1. 接口用途

此接口用于用户在商品详情页点击购买商品。

2. 参数校验

校验用户令牌是否为空，校验商品编号是否为空，校验收货地址编号是否为空。

3. 业务规则

校验 token 是否有效；根据商品编号查询商品是否存在及商品状态是否正常；根据收货地址编号查询地址信息是否存在；生成订单的同时往队列发送一条延时消息，延时时间为 24 小时，用于超时自动取消订单。

4. 请求方式

采用 POST 请求。

5. 请求报文

购买商品接口请求报文如表 23-1 所示。

表 23-1 购买商品接口请求报文说明

参数名	必需	类型	字节长度	说明
productId	是	int	11	商品编号
addressId	是	int	11	收货地址编号

6. 请求示例

```
{
"productId":1,
"addressId":1
}
```

7. 编写程序

（1）OrderController 类，订单相关的接口控制层，placeOrder 方法用于接收客户端购买商品请求。主要包含校验用户令牌、商品编号、收货地址编号是否为空，然后调用 service 层业务，并返回报文信息给客户端。

```
/**
 * 下单购买商品
 * @param vo
 * @return
 */
@RequestMapping(value = "placeOrder", method = RequestMethod.POST,
produces = MediaType.APPLICATION_JSON_UTF8_VALUE)
public ResponseBO placeOrder(@RequestBody OrderVO vo) {
    // 校验token是否为空
    String token = checkToken();
    Assertion.isPositive(vo.getProductId(), "商品编号不能为空");
    Assertion.isPositive(vo.getAddressId(), "收货地址编号不能为空");
    vo.setToken(token);
    return ResponseBO.success(orderService.placeOrder(vo));
}
```

（2）OrderServiceImpl 类，placeOrder 方法用于购买商品业务。校验 token 有效性，若校验不通过，则返回相应的错误信息。否则继续判断商品信息、地址信息是否存在及状态是否正常，若不存在，则返回相应的错误信息。生成订单后发送延时队列，用于订单超过 24 小时未处理自动取消，并将订单有效时间存入 redis，用于查询返回给客户端展示倒计时。最后将订单信息组装并返回。

```java
    /**
     * 购买商品
     * @param vo
     * @return
     */
    @Override
    public OrderBO placeOrder(OrderVO vo) {
        try {
            // 校验 token 是否有效
            User user = userService.checkToken(vo.getToken());
            // 校验商品是否存在
            ProductBO productBO = productMapper.selectProductInfoAndTradeStatus(vo.getProductId());
            Assertion.notNull(productBO, "商品信息不存在");
            Assertion.isTrue(2 == productBO.getTradeStatus(), "该商品正在交易或已卖出");
            Address address = addressMapper.selectByPrimaryKey(vo.getAddressId());
            Assertion.notNull(address, "收货信息不存在");
            Assertion.isTrue(1 == address.getStatus(), "收货信息已被删除");
            // 设置 redis 锁，防止商品被重复交易
            long productLock = RedisUtils.setNx(String.format(RedisConstants.PRODUCT_ID, vo.getProductId()), 60, "1");
            if (productLock == 0) {
                log.info("商品正在被交易,{}", vo.getProductId());
                throw new BusinessException(-1, "商品正在交易");
            }
            Date date = new Date();
            // 生成待支付订单
            Order order = new Order();
            order.setBuyingUserId(user.getUserId());
            order.setCreateTime(date);
            order.setOrderAmount(productBO.getProductPrice());
            order.setProductId(productBO.getId());
            order.setUpdateTime(date);
            String consigneeAddress = address.getProvince() + " " +
```

```
address.getCity() + address.getDistrict() + address.getStreet()
            + " " + address.getAddressDetail();
        order.setConsigneeAddress(consigneeAddress);
        order.setConsigneeMobile(address.getConsigneeMobile());
        order.setConsigneeName(address.getConsigneeName());
        orderMapper.insertSelective(order);
        int orderId = order.getOrderId();
        // 发送延时队列，用于订单超过 24 小时未处理时自动取消
        directMqSender.sendTtlMessage(RabbitMqEnum.RoutingEnum.
SECONDARY_ORDER_DEAD_LETTER_ROUTING, orderId + "");
        // 订单有效时长写入 redis
        String key = String.format(RedisConstants.ORDER_ID, orderId);
        RedisUtils.setEx(key, ttl / 1000, orderId + "");
        // 组装返回信息
        OrderBO bo = new OrderBO();
        bo.setOrderId(orderId);
        bo.setCreateTime(date);
        bo.setOrderAmount(order.getOrderAmount());
        bo.setPayStatus(1);
        bo.setTradeStatus(1);
        bo.setProductDesc(productBO.getProductDesc());
        bo.setProductImgs(productBO.getProductImgs());
        bo.setBuyerUserName(user.getUserName());
        bo.setExpiredTime((long) ttl);
        return bo;
    } catch (Exception e) {
        // 发生异常，删除 redis 锁
        RedisUtils.del(String.format(RedisConstants.PRODUCT_ID, vo.
getProductId()));
        log.error("购买商品发生异常", e);
    }
    return null;
}
```

（3）入参实体 OrderVO，对应客户端请求报文。

```
package com.xpwu.secondary.vo;

import lombok.Getter;
```

```java
import lombok.Setter;
import lombok.ToString;

/**
 * Created by IntelliJ IDEA.
 *
 * @author:caoxue
 * @date:2019/9/15 15:25
 * @description: 下单接口入参
 * @version:1.0
 */
@Setter
@Getter
@ToString(callSuper = true)
public class OrderVO extends BaseVO {

    private static final long serialVersionUID = 7569023099860452795L;

    /**
     * 商品编号
     */
    private Integer productId;

    /**
     * 收货地址编号
     */
    private Integer addressId;

}
```

（4）返回报文实体 OrderBO，下单成功后返回信息实体。

```java
package com.xpwu.secondary.bo;

import lombok.Getter;
import lombok.Setter;
import lombok.ToString;

import java.io.Serializable;
import java.math.BigDecimal;
```

```java
import java.util.Date;

/**
 * Created by IntelliJ IDEA.
 *
 * @author:caoxue
 * @date:2019/9/21 16:17
 * @description:下单成功后返回信息
 * @version:1.0
 */
@Setter
@Getter
@ToString(callSuper = true)
public class OrderBO implements Serializable {
    private static final long serialVersionUID = 12607133368186600984L;

    /**
     * 订单编号
     */
    private Integer orderId;

    /**
     * 订单金额
     */
    private BigDecimal orderAmount;

    /**
     * 支付状态
     */
    private Integer payStatus;

    /**
     * 交易状态
     */
    private Integer tradeStatus;

    /**
     * 买家昵称
     */
    private String buyerUserName;
```

```java
    /**
     * 下单时间
     */
    private Date createTime;

    /**
     * 商品描述
     */
    private String productDesc;

    /**
     * 商品图片
     */
    private String productImgs;

    /**
     * 收货人手机号
     */
    private String consigneeMobile;

    /**
     * 收货人姓名
     */
    private String consigneeName;

    /**
     * 收货人地址
     */
    private String consigneeAddress;

    /**
     * 订单剩余有效时间（s）
     */
    private Long expiredTime;

}
```

8. 返回报文

购买商品接口返回报文如表 23-2 所示。

表 23-2 购买商品接口返回报文说明

参数名	必需	类型	字节长度	说明
orderId	是	int	11	订单编号
orderAmount	是	double	15	订单金额
payStatus	是	int	2	支付状态：1——待付款；2——付款中；3——已付款；4——付款失败
tradeStatus	是	int	12	交易状态：1——已下单；2——已取消；3——已结算
buyerUserName	是	string	50	买家昵称
createTime	是	long	13	下单时间
productDesc	是	string	200	商品描述
productImgs	是	string	2000	商品图片
consigneeMobile	是	string	20	收货人手机号
consigneeName	是	string	50	收货人姓名
consigneeAddress	是	string	500	收货人地址
expiredTime	是	long	13	订单剩余有效时间（s）

9. 返回示例

```
{
"code":0,
"message":"success",
"requestId":null,
"timestamp":1567301655000,
"result":{
  "orderId":1,
  "orderAmount":100,
  "payStatus":1,
  "tradeStatus":1,
  "buyerUserName":"xxx",
  "createTime":1567301655000,
  "productDesc":"xxx",
  "productImgs":"xxx",
  "consigneeMobile":"13888888888",
  "consigneeName":"xxx",
```

```
"consigneeAddress":"xxx",
"expiredTime":10000
}
```

23.2 获取订单详情接口

1. 接口用途

此接口用于用户在订单列表页点击查询详情信息。

2. 参数校验

校验用户令牌是否为空，校验订单编号是否为空。

3. 业务规则

校验 token 是否有效，根据订单编号查询订单是否存在，从 redis 中取出订单剩余有效时间。

4. 请求方式

采用 GET 请求。

5. 请求报文

获取订单详情接口请求报文如表 23-3 所示。

表 23-3 获取订单详情接口请求报文说明

参数名	必需	类型	字节长度	说明
orderId	是	int	11	订单编号

6. 请求示例

GET 请求，参数直接放在 URL 后面。

7. 编写程序

（1）OrderController 类，订单相关的接口控制层，getOrderDetails 方法用于查询订单详情数据。主要包含校验用户令牌、订单编号是否为空，然后调用 service 层业务，并返回报文信息给客户端。

```
/**
 * 查询订单详情
 * @param vo
 * @return
```

```java
     */
    @RequestMapping(value = "getOrderDetails", method = RequestMethod.GET)
    public ResponseBO getOrderDetails(CommonOrderVO vo) {
        // 校验token是否为空
        String token = checkToken();
        Assertion.isPositive(vo.getOrderId(), "订单编号不能为空");
        vo.setToken(token);
        return ResponseBO.success(orderService.getOrderDetail(vo));
    }
```

（2）OrderServiceImpl 类，getOrderDetails 方法用于查询商品详情业务。首先校验 token 有效性，若校验不通过，则返回相应的错误信息。否则继续查询订单信息是否存在，若不存在，则返回相应的错误信息。然后从 redis 中取出订单剩余有效时间。最后将订单信息组装并返回。

```java
    /**
     * 查询订单详情
     * @param vo
     * @return
     */
    @Override
    public OrderBO getOrderDetails(CommonOrderVO vo) {
        // 校验token
        userService.checkToken(vo.getToken());
        // 查询订单信息
        OrderBO bo = orderMapper.selectOrderDetailByOrderId(vo.getOrderId());
        Assertion.notNull(bo, "订单信息不存在");
        // 从redis中取出订单剩余有效时间
        String key = String.format(RedisConstants.ORDER_ID, vo.getOrderId());
        long expiredTime = RedisUtils.pttl(key);
        log.info("expiredTime:" + expiredTime);
        bo.setExpiredTime(Detect.isPositive(expiredTime) ? expiredTime / 1000 :0);
        return bo;
    }
```

（3）入参实体 CommonOrderVO，对应客户端请求报文。

```java
package com.xpwu.secondary.vo;

import lombok.Getter;
import lombok.Setter;
import lombok.ToString;

/**
 * Created by IntelliJ IDEA.
 *
 * @author:caoxue
 * @date:2019/10/14 8:46
 * @description:订单通用入参
 * @version:1.0
 */
@Setter
@Getter
@ToString(callSuper = true)
public class CommonOrderVO extends BaseVO {

    private static final long serialVersionUID = -9078025283345783996L;
    /**
     * 订单编号
     */
    private Integer orderId;

}
```

（4）返回报文实体 OrderBO，获取订单详情成功后成功后返回信息实体。

```java
package com.xpwu.secondary.bo;

import lombok.Getter;
import lombok.Setter;
import lombok.ToString;

import java.io.Serializable;
import java.math.BigDecimal;
```

```java
import java.util.Date;

/**
 * Created by IntelliJ IDEA.
 *
 * @author:caoxue
 * @date:2019/9/21 16:17
 * @description:下单成功后返回信息
 * @version:1.0
 */
@Setter
@Getter
@ToString(callSuper = true)
public class OrderBO implements Serializable {
    private static final long serialVersionUID = 1260713368186600984L;

    /**
     * 订单编号
     */
    private Integer orderId;

    /**
     * 订单金额
     */
    private BigDecimal orderAmount;

    /**
     * 支付状态
     */
    private Integer payStatus;

    /**
     * 交易状态
     */
    private Integer tradeStatus;

    /**
     * 买家昵称
     */
    private String buyerUserName;
```

```java
    /**
     * 下单时间
     */
    private Date createTime;

    /**
     * 商品描述
     */
    private String productDesc;

    /**
     * 商品图片
     */
    private String productImgs;

    /**
     * 收货人手机号
     */
    private String consigneeMobile;

    /**
     * 收货人姓名
     */
    private String consigneeName;

    /**
     * 收货人地址
     */
    private String consigneeAddress;

    /**
     * 订单剩余有效时间（s）
     */
    private Long expiredTime;

}
```

8. 返回报文

获取订单详情接口返回报文如表 23-4 所示。

表 23-4　获取订单详情接口返回报文说明

参数名	必需	类型	字节长度	说明
orderId	是	int	11	订单编号
orderAmount	是	double	15	订单金额
payStatus	是	int	2	支付状态：1——待付款；2——付款中；3——已付款；4——付款失败
tradeStatus	是	int	12	交易状态：1——已下单；2——已取消；3——已结算
buyerUserName	是	string	50	买家昵称
createTime	是	long	13	下单时间
productDesc	是	String	200	商品描述
productImgs	是	string	2000	商品图片
consigneeMobile	是	string	20	收货人手机号
consigneeName	是	string	50	收货人姓名
consigneeAddress	是	string	500	收货人地址
expiredTime	是	long	13	订单剩余有效时间（s）

9. 返回示例

```
{
"code":0,
"message":"success",
"requestId":null,
"timestamp":1567301655000,
"result":{
  "orderId":1,
  "orderAmount":100,
  "payStatus":1,
  "tradeStatus":1,
  "buyerUserName":"xxx",
  "createTime":1567301655000,
  "productDesc":"xxx",
  "productImgs":"xxx",
  "consigneeMobile":"13888888888",
```

```
"consigneeName":"xxx",
"consigneeAddress":"xxx",
"expiredTime":10000
}
```

23.3 取消订单接口

1. 接口用途

此接口用于用户下单后在未超时、未付款的情况下主动取消订单。

2. 参数校验

校验用户令牌是否为空,校验订单编号是否为空。

3. 业务规则

校验 token 是否有效,根据订单编号查询订单是否存在,校验订单状态是否正常。

4. 请求方式

采用 POST 请求。

5. 请求报文

取消订单接口请求报文如表 23-5 所示。

表 23-5 取消订单接口请求报文说明

参数名	必需	类型	字节长度	说明
orderId	是	int	11	订单编号

6. 请求示例

```
{
"orderId":1
}
```

7. 编写程序

(1) OrderController 类,订单相关的接口控制层,cancelOrder 方法用于用户取消订单操作。主要包含校验用户令牌、订单编号是否为空,然后调用 service 层业务,并返回报文信息给客户端。

```java
/**
 * 取消订单
 * @param vo
 * @return
 */
@RequestMapping(value = "cancelOrder", method = RequestMethod.POST)
public ResponseBO cancelOrder(@RequestBody CommonOrderVO vo) {
    // 校验token是否为空
    String token = checkToken();
    Assertion.isPositive(vo.getOrderId(), "订单编号不能为空");
    vo.setToken(token);
    orderService.cancelOrder(vo);
    return ResponseBO.success();
}
```

（2）OrderServiceImpl 类，cancelOrder 方法用于取消订单业务。校验 token 有效性，若校验不通过，则返回相应的错误信息。否则继续查询订单信息是否存在，若不存在，则返回相应的错误信息。如果订单交易状态为已取消 / 已结算或者支付状态为付款中 / 已付款，则抛出对应的异常信息。校验通过后将订单交易状态更新为 2（已取消），完成取消订单操作。

```java
/**
 * 取消订单
 * @param vo
 */
@Override
public void cancelOrder(CommonOrderVO vo) {
    // 校验token
    userService.checkToken(vo.getToken());
    // 校验订单是否存在
    Order order = orderMapper.selectByPrimaryKey(vo.getOrderId());
    Assertion.notNull(order, "订单不存在");
    Assertion.isTrue(1 == order.getTradeStatus(), "订单已取消或已结算，请勿重复操作");
    Assertion.isTrue(1 == order.getPayStatus() || 4 == order.getPayStatus(), "订单正在付款或已付款，无法取消");
    Order update = new Order();
    update.setOrderId(vo.getOrderId());
```

```java
        update.setTradeStatus(2);
        update.setUpdateTime(new Date());
        orderMapper.updateByPrimaryKeySelective(update);
    }
```

（3）入参实体 CommonOrderVO，对应客户端请求报文。

```java
package com.xpwu.secondary.vo;

import lombok.Getter;
import lombok.Setter;
import lombok.ToString;

/**
 * Created by IntelliJ IDEA.
 *
 * @author:caoxue
 * @date:2019/10/14 8:46
 * @description:订单通用入参
 * @version:1.0
 */
@Setter
@Getter
@ToString(callSuper = true)
public class CommonOrderVO extends BaseVO {

    private static final long serialVersionUID = -9078025283345783996L;
    /**
     * 订单编号
     */
    private Integer orderId;

}
```

8. 返回报文及示例

参见公共返回示例。

23.4 支付宝 WAP 支付接口

(1) 登录支付宝开放平台，点击【网页 & 移动应用列表】，如图 23-1 所示。

图 23-1

(2) 点击【自定义接入】，准备创建应用，如图 23-2 所示。

图 23-2

(3) 填写应用名称，上传应用图标，应用类型选择【网页应用】，网址 url 与应用简介可以不填,点击【确认创建】,如图 23-3 所示。然后等待支付宝审核,一般需要 1 ~ 2 个工作日。

图 23-3

(4)审核通过之后,进入图 23-2 所示的页面,我的应用列表会出现一条记录,点击【查看详情】,如图 23-4 所示。

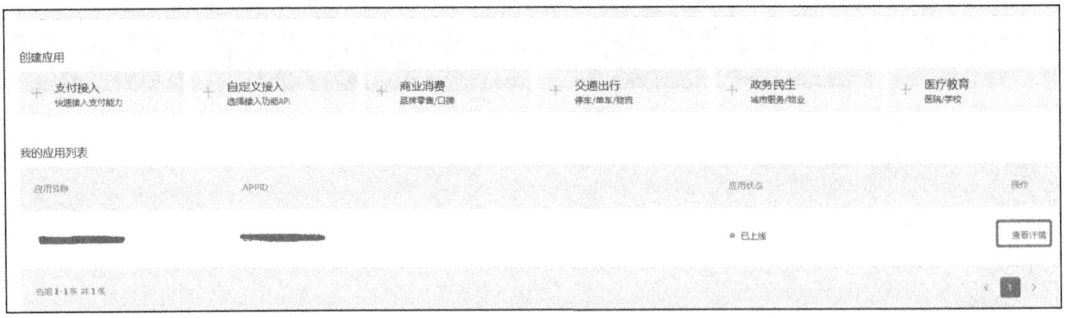

图 23-4

(5)接下来要进行签约,点击【签约】,如图 23-5 所示。
(6)填写相应的信息,等待支付宝审核,如图 23-6 所示。
(7)审核过程中可以先进行一些前期准备,比如设置接口加签方式,点击【应用信息】，设置接口加签方式，如图 23-7 所示。

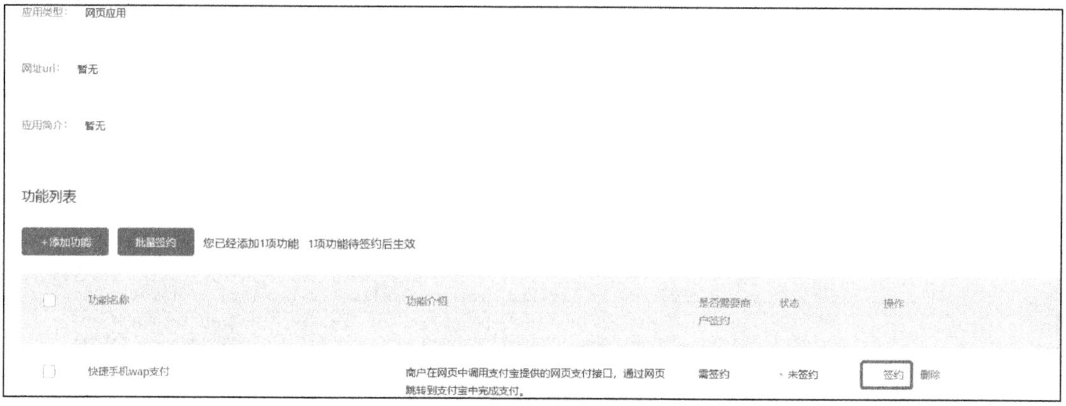

图 23-5

图 23-6

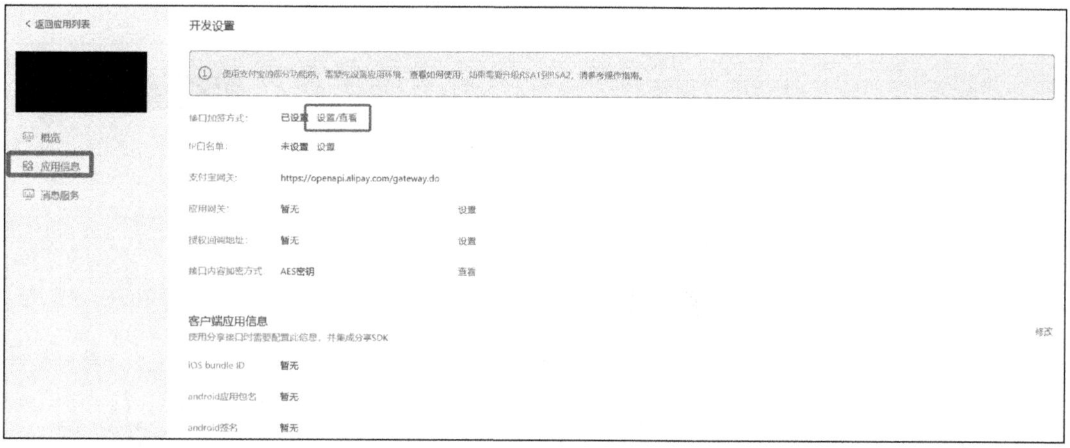

图 23-7

（8）这个时候需要填写应用公钥，点击【支付宝秘钥生成器】，如图 23-8 所示。

图 23-8

（9）根据文档中心的文档提示，下载支付宝开放平台开发助手并生成公钥私钥，如图 23-9 所示。将公钥复制到图 23-8 中填写公钥字符的文本框内，保存设置。同时将私钥保存在本地，供后续使用。

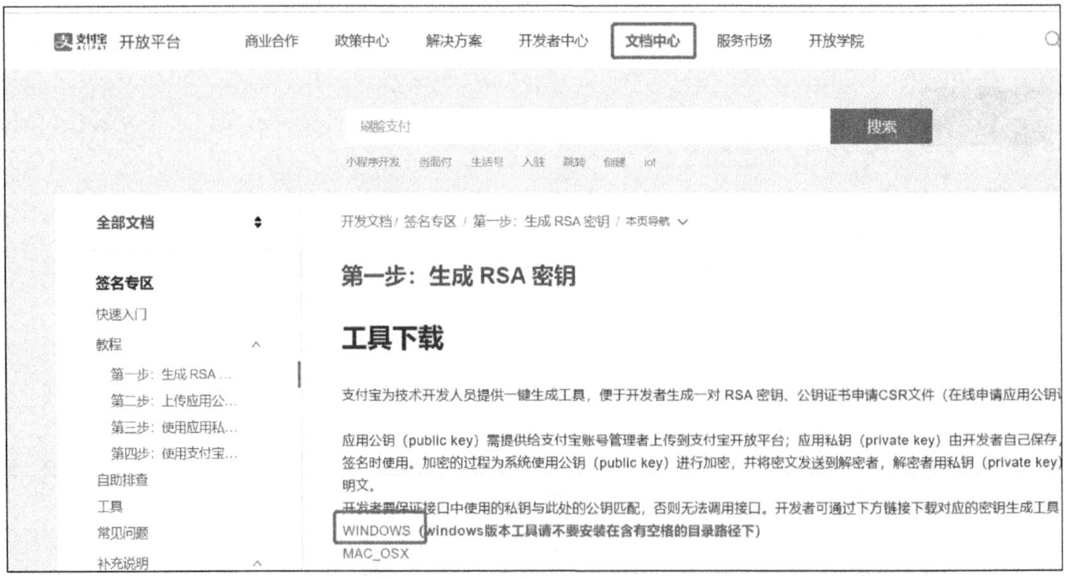

图 23-9

(10)公钥配置完成,如图 23-10 所示。

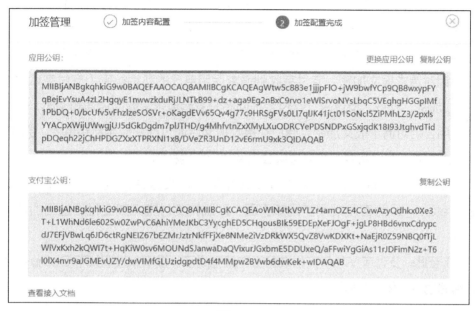

图 23-10

(11)下载 SDK&Demo,第一次对接,下载官方 Demo 比较容易上手。在文档中心搜索开发文档→手机网站支付,点击【SDK&Demo】,选择 JAVA 版下载,如图 23-11 所示。

图 23-11

（12）将下载的 Demo 导入开发工具中，配置 AlipayConfig 文件。注意 RSA_PRIVATE_KEY 为商户自己生成的私钥，ALIPAY_PUBLIC_KEY 为支付宝公钥，并非商户公钥。

这里解释一下公钥加密、私钥解密，这个很容易理解。商户在请求支付宝时，通过支付宝公钥对请求报文加密，支付宝收到请求后用自己的私钥解密（支付宝私钥内部配置），完成加签、验签。反过来，支付宝在向商户发起请求（如回调通知）时，通过商户上传的应用公钥对请求报文加密，商户在接收报文后通过商户应用私钥解密，完成加签、验签。综上所述，商户需要配置支付宝公钥及商户应用私钥。

再补充一点，notify_url 为后端异步通知地址，服务端需要开发一个 API，并设置进去；return_url 为客户端同步通知页面，即跳转支付宝完成支付后返回的前端页面地址。

（13）至此，支付宝 WAP 支付准备工作就完成了。接下来按照如下 Demo 接入支付请求，后面也有具体的对接。

```jsp
<%
/* *
 * 功能：支付宝手机网站支付接口 (alipay.trade.wap.pay) 接口调试入口页面
 * 版本：2.0
 * 修改日期：2016-11-01
 * 说明：
 * 以下代码只是为了方便商户测试而提供的样例代码，商户可以根据自己网站的需要，按
照技术文档编写，并非一定要使用该代码。
 * 请确保项目文件有可写权限，不然无法输出日志。
 */
%>
<%
if(request.getParameter("WIDout_trade_no")!=null){
    // 商户订单号，商户网站订单系统中唯一订单号，必填
    String out_trade_no = new String(request.getParameter("WIDout_trade_no").getBytes("ISO-8859-1"),"UTF-8");
    // 订单名称，必填
    String subject = new String(request.getParameter("WIDsubject").getBytes("ISO-8859-1"),"UTF-8");
    System.out.println(subject);
    // 付款金额，必填
    String total_amount=new String(request.getParameter("WIDtotal_
```

```java
amount").getBytes("ISO-8859-1"),"UTF-8");
        // 商品描述，可留空
        String body = new String(request.getParameter("WIDbody").getBytes
("ISO-8859-1"),"UTF-8");
        // 超时时间，可留空
     String timeout_express="2m";
        // 销售商品码，必填
        String product_code="QUICK_WAP_WAY";
        /***********************/
        // SDK 公共请求类，包含公共请求参数，以及封装了签名与验签，开发者无须关注签
名与验签
        // 调用 RSA 签名方式
        AlipayClient client = new DefaultAlipayClient(AlipayConfig.URL,
AlipayConfig.APPID, AlipayConfig.RSA_PRIVATE_KEY, AlipayConfig.FORMAT,
AlipayConfig.CHARSET, AlipayConfig.ALIPAY_PUBLIC_KEY,AlipayConfig.SIGNTYPE);
        AlipayTradeWapPayRequest alipay_request=new AlipayTradeWapPayRequest();

        // 封装请求支付信息
        AlipayTradeWapPayModel model=new AlipayTradeWapPayModel();
        model.setOutTradeNo(out_trade_no);
        model.setSubject(subject);
        model.setTotalAmount(total_amount);
        model.setBody(body);
        model.setTimeoutExpress(timeout_express);
        model.setProductCode(product_code);
        alipay_request.setBizModel(model);
        // 设置异步通知地址
        alipay_request.setNotifyUrl(AlipayConfig.notify_url);
        // 设置同步地址
        alipay_request.setReturnUrl(AlipayConfig.return_url);

        // 生成 form 表单
        String form = "";
        try {
            // 调用 SDK 生成表单
            form = client.pageExecute(alipay_request).getBody();
            response.setContentType("text/html;charset=" + AlipayConfig.
CHARSET);
            response.getWriter().write(form);// 直接将完整的表单html输出到页面
            response.getWriter().flush();
```

```
        response.getWriter().close();
    } catch (AlipayApiException e) {
        // TODO Auto-generated catch block
        e.printStackTrace();
    }
}
%>
```

23.5 支付宝支付接口

1. 接口用途

此接口用于用户下单后通过支付宝对订单进行付款。

2. 参数校验

校验用户令牌是否为空,校验订单编号是否为空。

3. 业务规则

校验 token 是否有效,根据订单编号查询订单是否存在,校验订单状态是否正常。

4. 请求方式

采用 GET 请求。

5. 请求报文

支付宝支付接口请求报文如表 23-6 所示。

表 23-6 支付宝支付接口请求报文说明

参数名	必需	类型	字节长度	说明
orderId	是	int	11	订单编号
token	是	string	32	用户令牌

6. 请求示例

```
{
"orderId":1,
"token":"xxx"
}
```

7. 编写程序

(1) AlipayController 类,支付相关的接口控制层,alipay 方法用于用户下单后进行支付宝 WAP 支付。主要包含校验用户令牌、订单编号是否为空，然后调用 service 层业务生成表单，并返回给客户端，客户端自动跳转到支付宝进行支付。

```java
/**
 * 支付宝网页支付
 *
 * @param vo
 * @param response
 */
@RequestMapping(value = "alipay", method = {RequestMethod.POST, RequestMethod.GET})
public void alipay(AlipayVO vo, HttpServletResponse response) {
    // 校验 token 是否为空
    Assertion.isPositive(vo.getOrderId(), "订单号不能为空");
    Assertion.notEmpty(vo.getToken(), "token 不能为空");
    String form = alipayService.alipay(vo);
    try {
        response.setContentType("text/html;charset=" + AlipayConfig.CHARSET);
        // 直接将完整的表单 html 输出到页面
        PrintWriter out = response.getWriter();
        out.write(form);
        out.flush();
        out.close();
    } catch (Exception e) {
        log.error("支付宝调用支付接口异常", e);
    }
}
```

（2）AlipayServiceImpl 类，alipay 方法用于支付宝 WAP 支付业务。首先校验 token 有效性，若校验不通过，则返回相应的错误信息。否则继续查询订单信息是否存在，若不存在，则返回相应的错误信息。如果订单交易状态为已取消 / 已结算或者支付状态为付款中 / 已付款，则抛出对应的异常信息。校验通过后，组装支付宝 WAP 支付请求报文，并调用支付宝 SDK 生成表单信息，同时修改订单支付状态为支付中并生成支付信息，最后将支付宝表单信息返回。

```
/**
 * 支付宝支付
```

```java
     * @param vo
     * @return
     */
    @Override
    @Transactional(rollbackFor = Exception.class, isolation = Isolation.READ_COMMITTED)
    public String alipay(AlipayVO vo) {
        User user = userService.checkToken(vo.getToken());
        // 查询订单状态
        Order order = orderMapper.selectByPrimaryKey(vo.getOrderId());
        Assertion.notNull(order, "订单不存在");
        Assertion.isTrue(1 == order.getTradeStatus(), "订单已取消或已结算");
        Assertion.isTrue(1 == order.getPayStatus() || 4 == order.getPayStatus(), "订单正在付款或已付款");
        // 商户订单号，商户网站订单系统中唯一订单号，必填
        String outTradeNo = System.currentTimeMillis() + "";
        // 订单名称，必填
        String subject = "手机网站支付测试商品";
        // 付款金额，必填
        String totalAmount = String.valueOf(order.getOrderAmount());
        // 商品描述，可留空
        String body = "购买测试商品";
        // 超时时间，可留空
        String timeoutExpress = "2m";
        // 销售商品码，必填
        String productCode = "QUICK_WAP_WAY";
        // SDK 公共请求类，包含公共请求参数，以及封装了签名与验签，开发者无需关注签名与验签
        // 调用 RSA 签名方式
        AlipayClient client = new DefaultAlipayClient(AlipayConfig.URL, AlipayConfig.APPID, AlipayConfig.RSA_PRIVATE_KEY, AlipayConfig.FORMAT, AlipayConfig.CHARSET, AlipayConfig.ALIPAY_PUBLIC_KEY, AlipayConfig.SIGNTYPE);
        AlipayTradeWapPayRequest alipayRequest = new AlipayTradeWapPayRequest();
        // 封装请求支付信息
        AlipayTradeWapPayModel model = new AlipayTradeWapPayModel();
        model.setOutTradeNo(outTradeNo);
        model.setSubject(subject);
        model.setTotalAmount(totalAmount);
        model.setBody(body);
        model.setTimeoutExpress(timeoutExpress);
```

```java
            model.setProductCode(productCode);
            alipayRequest.setBizModel(model);
            // 设置后端异步通知地址
            alipayRequest.setNotifyUrl(AlipayConfig.NOTIFY_URL);
            // 设置前端同步返回地址
            alipayRequest.setReturnUrl(AlipayConfig.RETURN_URL);
            log.info("alipayRequest:{}", JSON.toJSONString(alipayRequest));
            // 生成 form 表单
            String form = "";
            try {
                // 调用 SDK 生成表单
                form = client.pageExecute(alipayRequest).getBody();
                log.info("form:{}", form);
            } catch (AlipayApiException e) {
                log.error("调用支付宝支付接口异常", e);
            }
            // 修改订单支付状态
            Order update = new Order();
            Date date = new Date();
            update.setOrderId(order.getOrderId());
            update.setUpdateTime(date);
            update.setPayTime(date);
            update.setPayStatus(2);
            orderMapper.updateByPrimaryKeySelective(update);
            // 生成支付信息
            PaymentLog paymentLog = this.buildPaymentLog(alipayRequest, outTradeNo, totalAmount, user.getUserId(), vo.getOrderId());
            paymentLogMapper.insertSelective(paymentLog);
            return form;
        }
```

（3）入参实体 AlipayVO，对应客户端请求报文。

```java
package com.xpwu.secondary.vo;

import lombok.Getter;
import lombok.Setter;
import lombok.ToString;

/**
```

```
 * Created by IntelliJ IDEA.
 *
 * @author:caoxue
 * @date:2019/9/16 20:36
 * @description:支付宝支付请求报文
 * @version:1.0
 */
@Setter
@Getter
@ToString(callSuper = true)
public class AlipayVO extends BaseVO {

    private static final long serialVersionUID = 761808750973621005L;
    /**
     * 订单编号
     */
    private Integer orderId;

}
```

8. 返回报文及示例

返回支付宝付款页面。

23.6 支付宝支付回调接口

1. 接口用途

完成支付后，支付宝向服务端发起异步回调通知，服务端用于修改订单信息及支付信息。

2. 参数校验

支付宝验签。

3. 业务规则

支付宝发起支付成功回调，通知服务端支付结果及支付信息。服务端修改订单、支付信息。

4. 请求方式

采用 GET 请求。

5. 请求报文

见支付宝官网文档。

6. 请求示例

```
{
"orderId":1
}
```

7. 编写程序

（1）AlipayController 类，支付相关接口控制层，alipayNotify 方法用于接收支付宝支付回调请求。直接调用 service 层业务，并返回报文信息给支付宝。

```java
/**
 * 支付宝支付回调 notify_url
 *
 * @param request
 * @param response
 */
@RequestMapping(value = "alipayNotify", method = RequestMethod.POST)
public void alipayNotify(HttpServletRequest request, HttpServletResponse response) {
    try {
        PrintWriter out = response.getWriter();
        String res = alipayService.alipayNotify(request);
        log.info("支付回调 res:" + res);
        out.println(res);
        out.flush();
        out.close();
    } catch (IOException e) {
        log.error("支付宝回调异常", e);
    }
}
```

（2）AlipayServiceImpl 类，alipayNotify 方法为支付宝支付回调功能。获取支付宝传入参数后记录回调流水并进行签名验证，若验证失败，直接返回"fail"给支付宝；若验证成功，则根据传入的交易状态修改订单表及支付流水表中的支付状态、交易状态等信息。只要签名验证成功，就返回"success"，回调业务处理完成。

```java
    /**
     * 支付宝回调
     * @param request
     * @return
     */
    @Override
    @Transactional(rollbackFor = Exception.class, isolation = Isolation.READ_COMMITTED)
    public String alipayNotify(HttpServletRequest request) {
        // 获取支付宝的反馈信息
        Map<String, String> params = new HashMap<>(16);
        Map requestParams = request.getParameterMap();
        try {
            for (Object o :requestParams.keySet()) {
                String name = (String) o;
                String[] values = (String[]) requestParams.get(name);
                String valueStr = "";
                for (int i = 0; i < values.length; i++) {
                    valueStr = (i == values.length - 1) ? valueStr + values[i]
                            :valueStr + values[i] + ",";
                }
                // 乱码解决,这段代码在出现乱码时使用。如果mysign和sign不相等也可以使用这段代码转化
                //valueStr = new String(valueStr.getBytes("ISO-8859-1"), "gbk")
                params.put(name, valueStr);
            }
            // 获取支付宝的通知返回参数,可参考技术文档中页面跳转同步通知参数列表(以下仅供参考)
            // 商户订单号
            String outTradeNo = new String(request.getParameter("out_trade_no").getBytes(StandardCharsets.ISO_8859_1), StandardCharsets.UTF_8);
            // 支付宝交易号
            String tradeNo = new String(request.getParameter("trade_no").getBytes(StandardCharsets.ISO_8859_1), StandardCharsets.UTF_8);
            // 交易状态
            String tradeStatus = new String(request.getParameter("trade_status").getBytes(StandardCharsets.ISO_8859_1), StandardCharsets.UTF_8);
```

```java
            // 记录回调流水
            try {
                PaymentNotifyLog notifyLog = this.buildPaymentNotifyLog
(outTradeNo, tradeNo, tradeStatus, JSON.toJSONString(params));
                paymentNotifyLogMapper.insertSelective(notifyLog);
            } catch (Exception e) {
                log.warn("支付宝回调流水记录异常", e);
            }
            // 计算得出通知验证结果
            boolean AlipaySignature.rsaCheckV1(Map<String, String> params, String publicKey, String charset, String sign_type)
            boolean verifyResult = AlipaySignature.
rsaCheckV1(params, AlipayConfig.ALIPAY_PUBLIC_KEY, AlipayConfig.
CHARSET, AlipayConfig.SIGNTYPE);
            // 验证成功
            if (verifyResult) {
                log.info("验签通过");
                // 请在这里加上商户的业务逻辑程序代码
                // 请根据您的业务逻辑来编写程序(以下代码仅供参考)
                if (AlipayConstants.TRADE_FINISHED.equals(tradeStatus)
                        || AlipayConstants.TRADE_SUCCESS.equals(tradeStatus)) {
                    try {
                        // 修改订单表、支付状态
                        this.updateOrderAndPaymentLogStatus(outTrade
No, params, tradeNo);
                    } catch (Exception e) {
                        log.error("支付回调状态修改异常", e);
                    }
                    // 判断该笔订单是否在商户网站中已经做过处理
                    // 如果没有做过处理,根据订单号(out_trade_no)在商户网站的
订单系统中查到该笔订单的详情,并执行商户的业务逻辑程序
                    // 请务必判断请求时的 total_fee、seller_id 与通知时获取的
total_fee、seller_id是否是一致的
                    // 如果有做过处理,不执行商户的业务逻辑程序
                    // 注意:
                    // 如果签约了可退款协议,退款日期超过可退款期限后(如3个月可
退款),支付宝系统发送该交易状态通知
                    // 如果没有签约可退款协议,那么付款完成后,支付宝系统发送该交
易状态通知
                    // TRADE_SUCCESS 交易成功: 即时到账高级版。交易成功,但卖
家可以执行退款操作进行退款,即该交易还没有彻底完成,卖家还可以修改这笔交易。所以支付宝
```

后续还会有至少一条推送（TRADE_FINISHED）
                    // TRADE_FINISHED 交易完成：即时到账普通版。普通版不支持支付完成后的退款操作
                }
                return "success";
            } else {
                log.info(" 验签失败 ");
            }
        } catch (Exception e) {
            log.error(" 支付宝notify异常 ", e);
        }
        return "fail";
    }
    /**
     * 修改支付状态
     * @param outTradeNo
     * @param params
     * @param tradeNo
     */
    private void updateOrderAndPaymentLogStatus(String outTradeNo, Map<String, String> params, String tradeNo) {
        log.info(" 开始修改订单、支付信息 ");
        Example example = new Example(PaymentLog.class);
        example.createCriteria().andEqualTo("tradeNo", outTradeNo);
        PaymentLog paymentLog = paymentLogMapper.selectOneByExample(example);
        if (paymentLog == null) {
            throw new RuntimeException(" 支付信息不存在 ");
        }
        PaymentLog update = new PaymentLog();
        update.setOutTradeNo(tradeNo);
        update.setTradeStatus(params.get("trade_status"));
        update.setActualPayAmount(new BigDecimal(params.get("buyer_pay_amount")));
        update.setCompletionTime(DateUtils.parseStrToDate(params.get("gmt_payment")));
        update.setId(paymentLog.getId());
        paymentLogMapper.updateByPrimaryKeySelective(update);
        log.info(" 支付信息修改完成 ");
        Order order = orderMapper.selectByPrimaryKey(paymentLog.getOrderId());
        if (null == order) {
```
```

```
            throw new RuntimeException("订单信息不存在");
        }
        if (!order.getOrderAmount().equals(new BigDecimal(params.get("total_amount")))) {
            throw new RuntimeException("订单金额不一致");
        }
        Order entity = new Order();
        entity.setPayStatus(3);
        entity.setTradeStatus(3);
        entity.setOutTradeNo(tradeNo);
        entity.setPayAmount(new BigDecimal(params.get("buyer_pay_amount")));
        entity.setCompletionTime(DateUtils.parseStrToDate(params.get("gmt_payment")));
        entity.setOrderId(order.getOrderId());
        orderMapper.updateByPrimaryKeySelective(entity);
        log.info("订单记录修改完成");
    }
```

8. 返回报文及示例

返回字符串"success"或"fail"。

第 24 章 个人中心模块接口

【本章导读】

◎ 我的商品列表接口

◎ 删除我的商品接口

◎ 校验旧手机号接口

◎ 绑定新手机号接口

◎ 查询商品数量接口

◎ 收货地址列表接口

◎ 新增收货地址接口

◎ 修改收货地址接口

◎ 删除收货地址接口

24.1 我的商品列表接口

1. 接口用途

此接口用于用户在个人中心模块查询我发布的商品列表、我卖出的商品列表、我买到的商品列表数据。

2. 参数校验

校验 token 是否为空，类别是否为空且类别是否有效。当前页码不传默认为 1，每页条数不传默认为 10。

3. 业务规则

此接口需要用户登录后查询，需要校验 token 有效性。

4. 请求方式

采用 POST 请求。

5. 请求报文

我的商品列表接口请求报文如表 24-1 所示。

表 24-1 我的商品列表接口请求报文说明

参数名	必需	类型	字节长度	说明
type	是	int	2	查询类别：1——我发布的商品信息；2——我卖出的商品信息；3——我买到的商品信息
pageNum	否	int	6	当前页码，不传默认为 1
pageSize	否	int	6	每页条数，不传默认为 10

6. 请求示例

```
{
"type":1,
"pageNum":1,
"pageSize":10
}
```

7. 编写程序

（1）ProductController 类为商品相关的接口控制层，getMyProductList 方法用于查询我的商品列表数据。主要包含校验请求报文是否为空及参数值是否有效等，并调用 service 层业务，将报文返回给客户端。

```
/**
 * 获取商品列表
 * @param vo
 * 分类查询我发布的商品信息、我卖出的商品信息、我买到的商品信息
 * @return
 */
@RequestMapping(value = "getMyProductList", method = RequestMethod.POST)
public ResponseBO getMyProductList(@Valid @RequestBody MyProductSearchVO vo) {
    Assertion.isTrue(vo != null, "参数为空");
    // 校验 token 是否为空
    String token = checkToken();
    Assertion.isTrue(1 == vo.getType() || 2 == vo.getType() || 3 == vo.getType(), "type 类型错误");
```

```java
        vo.setToken(token);
        // 调用 service 层查询
        return ResponseBO.successPageInfo(productService.getMyProductList(vo));
    }

    /**
     * 分页数据封装
     *
     * @param list
     * 数据列表
     * @return
     */
    public static ResponseBO successPageInfo(List<?> list) {
        Map<String, Object> map = new HashMap<>(16);
        map.put("list", list);
        map.put("total", Math.toIntExact(new PageInfo<>(list).getTotal()));
        return new ResponseBO.Builder().result(map).build();
    }
```

（2）ProductServiceImpl 类，getMyProductList 方法用于我的商品列表查询业务。此接口需要登录后调用，首先校验 token 是否有效，然后开启分页查询，调用 Mapper 层获取数据。

```java
    /**
     * 获取商品列表
     * @param vo
     * 分类查询我发布的商品信息、我卖出的商品信息、我买到的商品信息
     * @return
     */
    @Override
    public List<ProductBO> getMyProductList(MyProductSearchVO vo) {
        // 校验 token 是否有效
        User user = userService.checkToken(vo.getToken());
        // 开启分页查询
        PageHelper.startPage(vo.getPageNum(), vo.getPageSize());
        return productMapper.selectMyProductList(vo.getType(), user.getUserId());
    }
```

（3）ProductMapper 类，自定义 selectMyProductList 接口查询。

```
/**
 * 我的商品查询
 * @param type
 * @param userId
 * @return
 */
List<ProductBO> selectMyProductList(@Param("type") Integer type, @Param("userId") String userId);
```

（4）ProductMapper.xml，此处属于扩展查询，需要另外写 SQL 语句连接查询。其中我发布的商品列表按照发布时间降序排列，我买到的商品列表和我卖出的商品列表按照交易完成时间降序排列，具体如下。

```
<!-- 我的商品查询 -->
<select id="selectMyProductList" parameterType="string" resultType="com.xpwu.secondary.vo.ProductBO">
    SELECT
    p.id,
    p.product_imgs productImgs,
    p.product_desc productDesc,
    p.product_price productPrice,
    p.product_type_id productTypeId,
    t.wantNum,
    pt.'name' productTypeName,
    p.publish_user_id publishUserId,
    u.user_name publishUserName,
    u.user_avatar publishUserAvatar,
    o.buying_user_id buyingUserId,
    o.trade_status tradeStatus,
    p.create_time createTime,
    o.completion_time tradeTime,
    o.order_id orderId,
    p.product_address productAddress
    FROM
    product p
    LEFT JOIN product_type pt ON p.product_type_id = pt.id
    LEFT JOIN (SELECT COUNT(1) wantNum,product_id FROM chat c GROUP
```

```
BY product_id) t ON p.id = t.product_id
    LEFT JOIN 'order' o ON p.id = o.product_id
    LEFT JOIN 'user' u ON p.publish_user_id = u.user_id
    <where>
      <if test="type != null and type == 1">
        and p.publish_user_id = #{userId}
        and (o.order_id IS NULL or o.trade_status = 2)
      </if>
      <if test="type != null and type == 2">
        and p.publish_user_id = #{userId}
        and o.selling_status = 1
        and o.trade_status in(1, 3)
      </if>
      <if test="type != null and type == 3">
        and o.buying_user_id = #{userId}
        and o.buying_status = 1
        and o.trade_status in(1, 3)
      </if>
    </where>
      group by p.id
      <if test="type != null and type == 1">
          order by p.create_time desc
      </if>
      <if test="type != null and type != 1">
          order by o.completion_time desc
      </if>
```

（5）入参实体 MyProductSearchVO，对应客户端请求报文。

```
package com.xpwu.secondary.vo;

import lombok.Getter;
import lombok.Setter;
import lombok.ToString;

import javax.validation.constraints.NotNull;

/**
 * Created by IntelliJ IDEA.
```

```
 *
 * @author:caoxue
 * @date:2019/8/8 16:42
 * @description: 我的商品查询接口入参
 * @version:1.0
 */
@Setter
@Getter
@ToString(callSuper = true)
public class MyProductSearchVO extends BaseVO {
    private static final long serialVersionUID = 1754850472175512237L;

    /**
     * 查询类别: 1——我发布的商品信息; 2——我卖出的商品信息; 3——我买到的商品信息
     */
    @NotNull(message = "type 不能为空 ")
    private Integer type;

}
```

（6）返回报文实体 ProductBO。

```
package com.xpwu.secondary.bo;

import lombok.Getter;
import lombok.Setter;
import lombok.ToString;

import java.io.Serializable;
import java.math.BigDecimal;
import java.util.Date;

/**
 * Created by IntelliJ IDEA.
 *
 * @author:caoxue
 * @date:2019/9/12 9:55
 * @description:
 * @version:1.0
```

```java
 */
@Setter
@Getter
@ToString(callSuper = true)
public class ProductBO implements Serializable {

    private static final long serialVersionUID = 8821081481707766726L;

    /**
     * 商品编号
     */
    private Integer id;

    /**
     * 商品图片，多个图片地址用英文逗号隔开
     */
    private String productImgs;

    /**
     * 商品描述
     */
    private String productDesc;

    /**
     * 商品价格
     */
    private BigDecimal productPrice;

    /**
     * 商品类别编号
     */
    private Integer productTypeId;

    /**
     * 商品类别名称
     */
    private String productTypeName;

    /**
     * 商品发布人用户编号
     */
```

```java
private String publishUserId;

/**
 * 商品发布人用户昵称
 */
private String publishUserName;

/**
 * 商品发布人用户头像
 */
private String publishUserAvatar;

/**
 * 购买人用户编号
 */
private String buyingUserId;

/**
 * 交易状态：1——未交易；2——交易中；3——交易成功；4——交易失败
 */
private Integer tradeStatus;

/**
 * 创建时间（发布时间）
 */
private Date createTime;

/**
 * 交易时间（购买时间）
 */
private Date tradeTime;

/**
 * 想要人数
 */
private Integer wantNum;

/**
 * 订单编号
 */
private Integer orderId;
```

```
    /**
     * 商品地址
     */
    private String productAddress;

}
```

8. 返回报文

我的商品列表接口返回报文如表 24-2 所示。

表 24-2　我的商品列表接口返回报文说明

参数名	必需	类型	字节长度	说明
id	是	int	11	商品编号
productImgs	是	string	2000	商品图片，多个图片地址用英文逗号隔开
productDesc	是	string	500	商品描述
productPrice	是	double	10	商品价格
productTypeId	是	Int	11	商品类别编号
productTypeName	是	string	100	商品类别名称
publishUserId	是	string	32	商品发布人用户编号
publishUserName	是	string	50	商品发布人用户昵称
publishUserAvatar	是	string	100	商品发布人用户头像
buyingUserId	否	string	32	购买人用户编号
createTime	是	long	13	创建时间（发布时间）
wantNum	是	int	11	想要人数（有多少人想要此商品）
tradeTime	否	long	13	交易时间（购买时间）
productAddress	是	string	500	商品地址
orderId	否	int	11	订单编号

9. 返回示例

```
{
"code":0,
"message":"success",
"requestId":null,
"timestamp":1567301655000,
```

```
"result":[
 {
  "id":1,
  "productImgs":"xxx",
  "productDesc":"全新手机",
  "productPrice":4000,
  "productTypeId":1,
  "productTypeName":"手机",
  "publishUserId":"xxx",
  "publishUserName":"xxx",
  "publishUserAvatar":"xxx",
  "buyingUserId":"xxx",
  "createTime":1567301655000,
  "wantNum":5,
  "tradeTime":null,
  "productAddress":"xxx",
  "orderId":1
 }
]
}
```

24.2 删除我的商品接口

1. 接口用途

此接口用于用户将个人中心的商品数据删除,可以删除我发布的、我卖出的及我买到的商品信息。

2. 参数校验

校验用户令牌是否为空,校验商品编号是否为空,校验点赞状态是否为空。

3. 业务规则

校验 token 是否有效;根据商品编号查询商品是否存在;当类型为 1 时,校验商品发布人是否为当前登录用户;校验商品状态及订单状态是否正常。

4. 请求方式

采用 POST 请求。

5. 请求报文

删除我的商品接口请求报文如表 24-3 所示。

表 24-3 删除我的商品接口请求报文说明

参数名	必需	类型	字节长度	说明
productId	是	int	11	商品编号
type	是	int	2	类型：1——删除我发布的商品信息；2——删除我卖出的商品信息；3——删除我买到的商品信息

6. 请求示例

```
{
"productId":1,
"type":1
}
```

7. 编写程序

（1）ProductController 类，商品相关的接口控制层，delProduct 方法用于接收客户端删除商品请求。主要包含校验用户令牌、商品编号、删除类型是否为空，校验类型参数是否有效，然后调用 service 层业务，并返回报文信息给客户端。

```
/**
 * 删除我发布的商品信息
 * @param vo
 * @return
 */
@RequestMapping(value = "delProduct", method = RequestMethod.POST)
public ResponseBO delProduct(@RequestBody DelProductVO vo) {
    // 校验token是否为空
    String token = checkToken();
    Assertion.isPositive(vo.getProductId(), "商品编号不能为空");
    Assertion.isPositive(vo.getType(), "type不能为空");
    // 校验类型是否有效
    Assertion.isTrue(1 == vo.getType() || 2 == vo.getType() || 3 == vo.getType(), "type参数异常");
    vo.setToken(token);
    // 调用删除业务
    productService.delProduct(vo);
    return ResponseBO.success();
}
```

（2）ProductServiceImpl 类，delProduct 方法用于删除商品业务。校验 token 有效性并返回用户实体，若校验不通过，则返回相应的错误信息。否则继续判断商品信息是否存在，若不存在，则返回相应的错误信息。

当 type 为 1，即删除我发布的商品信息时，校验商品发布人是否为当前登录人，只能删除本人发布的商品信息，交易中的商品也无法被删除。校验通过之后将商品删除。

当 type 为 2，即删除我卖出的商品信息时，校验当前商品状态是否为"已完成交易"，并且有订单编号，否则抛出对应的异常信息。校验通过后将订单表 selling_status 更新为 2（卖出状态为已删除，不会出现在我卖出的商品列表里）。

当 type 为 3，即删除我买到的商品信息时，同样校验商品状态是否为"已完成交易"，并且有订单编号，否则抛出对应的异常信息。校验通过后将订单表 buying_status 更新为 2（买到状态为已删除，不会出现在我买到的商品列表里）。

```java
/**
 * 删除商品信息
 * @param vo
 */
@Override
public void delProduct(DelProductVO vo) {
    // 校验 token 是否有效
    User user = userService.checkToken(vo.getToken());
    // 查询商品信息
    ProductBO product = productMapper.selectProductInfoAndTradeStatus(vo.getProductId());
    Assertion.notNull(product, "商品不存在");
    if (1 == vo.getType()) {
        // 删除我发布的商品信息
        Assertion.equals(user.getUserId(), product.getPublishUserId(), "只能删除自己发布的商品");
        Assertion.equals(product.getTradeStatus(), 2, "该商品正在交易，不能被删除");
        productMapper.deleteByPrimaryKey(vo.getProductId());
    } else {
        // 状态必须为已交易
        Assertion.equals(product.getTradeStatus(), 3, "商品状态异常");
        Assertion.isPositive(product.getOrderId(), "商品状态异常");
        Order order = new Order();
```

```java
            if (2 == vo.getType()) {
                // 删除我卖出的商品信息
                order.setSellingStatus(2);
            } else {
                // 删除我买到的商品信息
                order.setBuyingStatus(2);
            }
            order.setUpdateTime(new Date());
            order.setOrderId(product.getOrderId());
            orderMapper.updateByPrimaryKeySelective(order);
        }
    }
```

（3）入参实体 DelProductVO，对应客户端请求报文。

```java
package com.xpwu.secondary.vo;

import lombok.Getter;
import lombok.Setter;
import lombok.ToString;

/**
 * Created by IntelliJ IDEA.
 *
 * @author:caoxue
 * @date:2019/9/14 16:46
 * @description:删除商品入参
 * @version:1.0
 */
@Setter
@Getter
@ToString(callSuper = true)
public class DelProductVO extends BaseVO {
    private static final long serialVersionUID = 4626123382408864522L;

    /**
     * 商品编号
     */
    private Integer productId;
```

```
    /**
     * 删除类型：1——我发布的商品信息；2——我卖出的商品信息；3——我买到的商
品信息
     */
    private Integer type;
}
```

8. 返回报文及示例

参见公共返回示例。

24.3 校验旧手机号接口

1. 接口用途

此接口用于用户在更改手机号时进行第一步校验，校验旧手机号是否正确。

2. 参数校验

校验用户令牌、旧手机号及短信验证码是否为空。

3. 业务规则

校验 token 是否有效；校验客户端传入手机号是否与注册手机号一致；根据发送验证码类型及手机号组装 redis key，校验短信验证码是否正确。

4. 请求方式

采用 POST 请求。

5. 请求报文

校验旧手机号接口请求报文如表 24-4 所示。

表 24-4 校验旧手机号接口请求报文说明

参数名	必需	类型	字节长度	说明
mobile	是	string	20	旧手机号
verifyCode	是	string	2	短信验证码

6. 请求示例

{
"mobile":"13888888888",

```
"verifyCode":"123456"
}
```

7. 编写程序

（1）UserController 类，用户相关的接口控制层，verifyOldMobile 方法用于用户在更换手机号时进行第一步校验，检查旧手机号是否正确。主要包含校验用户令牌、旧手机号、短信验证码是否为空，然后调用 service 层业务，并返回报文信息给客户端。

```
/**
 * 校验旧手机号
 * @return
 */
@RequestMapping(value = "verifyOldMobile", method = RequestMethod.POST)
public ResponseBO verifyOldMobile(@RequestBody UpdateMobileVO vo) {
    // 校验token是否为空
    String token = checkToken();
    Assertion.notEmpty(vo.getMobile(), "旧手机号不能为空");
    Assertion.notEmpty(vo.getVerifyCode(), "验证码不能为空");
    vo.setToken(token);
    // 调用service层业务
    userService.verifyOldMobile(vo);
    return ResponseBO.success();
}
```

（2）UserServiceImpl 类，verifyOldMobile 方法用于校验旧手机号业务。首先校验 token 有效性，若校验不通过，则返回相应的错误信息，否则继续判断客户端传入手机号是否与注册手机号一致。然后从缓存中取出短信验证码，校验短信验证码是否正确，若校验不通过，则抛出对应的异常信息；若校验通过，则完成校验旧手机号业务。

```
/**
 * 校验旧手机号
 * @param vo
 */
@Override
public void verifyOldMobile(UpdateMobileVO vo) {
    // 校验token是否有效
    User user = this.checkToken(vo.getToken());
```

```java
        Assertion.equals(user.getMobile(), vo.getMobile(), "手机号必须与注册手机号一致");
        // 获取缓存中的验证码
        String code = RedisUtils.get(String.format(RedisConstants.OLD_MOBILE_VERIFY_CODE, vo.getMobile()));
        Assertion.notEmpty(code, "验证码已失效");
        Assertion.equals(vo.getVerifyCode(), code, "验证码错误");
    }
```

（3）入参实体 UpdateMobileVO，对应客户端请求报文。

```java
package com.xpwu.secondary.vo;

import lombok.Getter;
import lombok.Setter;
import lombok.ToString;

/**
 * Created by IntelliJ IDEA.
 *
 * @author:caoxue
 * @date:2019/9/14 16:19
 * @description:修改手机号入参
 * @version:1.0
 */
@Setter
@Getter
@ToString(callSuper = true)
public class UpdateMobileVO extends BaseVO {
    private static final long serialVersionUID = 4656015852611448064L;

    /**
     * 旧手机号
     */
    private String mobile;

    /**
     * 短信验证码
     */
    private String verifyCode;

}
```

8. 返回报文及示例

参见公共返回示例。

24.4 绑定新手机号接口

1. 接口用途

此接口用于用户在更改手机号时，通过旧手机号校验后，重新绑定新手机号。

2. 参数校验

校验用户令牌、新手机号及短信验证码是否为空。

3. 业务规则

校验 token 是否有效；校验客户端传入手机号是否与注册手机号一致；根据发送验证码类型及手机号组装 redis key，校验短信验证码是否正确。

4. 请求方式

采用 POST 请求。

5. 请求报文

绑定新手机号接口请求报文如表 24-5 所示。

表 24-5　绑定新手机号接口请求报文说明

参数名	必需	类型	字节长度	说明
mobile	是	string	20	新手机号
verifyCode	是	string	2	短信验证码

6. 请求示例

```
{
"mobile":"13888888888",
"verifyCode":"123456"
}
```

7. 编写程序

（1）UserController 类，用户相关的接口控制层，bindNewMobile 方法用于用户在更换手机号时绑定新手机号。主要包含校验用户令牌、新手机号、短信验证码是否为空，然后调用 service 层业务，并返回报文信息给客户端。

```java
/**
 * 绑定新手机号
 * @return
 */
@RequestMapping(value = "bindNewMobile", method = RequestMethod.POST)
public ResponseBO bindNewMobile(@RequestBody UpdateMobileVO vo) {
    // 校验 token 是否为空
    String token = checkToken();
    Assertion.notEmpty(vo.getMobile(), "新手机号不能为空");
    Assertion.notEmpty(vo.getVerifyCode(), "验证码不能为空");
    vo.setToken(token);
    // 调用 service 层绑定新手机号
    userService.bindNewMobile(vo);
    return ResponseBO.success();
}
```

（2）UserServiceImpl 类，bindNewMobile 方法用于绑定新手机号业务。首先校验 token 有效性，若校验不通过，则返回相应的错误信息。否则继续判断客户端传入手机号是否与原手机号是否一致，若一致，则抛出对应的异常信息。然后从缓存中取出短信验证码，校验短信验证码是否正确，若校验不通过，则抛出对应的异常信息。否则判断手机号是否已经注册，若已注册，则抛出对应的异常信息。最后更新用户信息及 token 信息，完成绑定新手机号业务。

```java
/**
 * 绑定新手机号
 * @param vo
 */
@Override
public void bindNewMobile(UpdateMobileVO vo) {
    // 校验 token 是否有效
    User user = this.checkToken(vo.getToken());
    Assertion.notEquals(user.getMobile(), vo.getMobile(), "不能与原手机号一样");
    // 获取缓存中的验证码
    String code = RedisUtils.get(String.format(RedisConstants.NEW_
```

```java
        MOBILE_VERIFY_CODE, vo.getMobile()));
        Assertion.notEmpty(code, "验证码已失效");
        Assertion.equals(vo.getVerifyCode(), code, "验证码错误");
        // 判断用户是否已经注册
        Example example = new Example(User.class);
        Example.Criteria criteria = example.createCriteria();
        criteria.andEqualTo("mobile", vo.getMobile());
        List<User> list = userMapper.selectByExample(example);
        Assertion.isEmpty(list, "手机号已注册");
        User update = new User();
        update.setMobile(vo.getMobile());
        update.setUpdateTime(new Date());
        update.setUserId(user.getUserId());
        BeanUtil.copyPropertiesIgnoreNull(update, user);
        this.updateUserInfo(update, user);
        // 更新token信息
        TokenBO bo = new TokenBO();
        bo.setMobile(vo.getMobile());
        bo.setUserId(user.getUserId());
        int expired = (int) ((user.getTokenExpired() - System.currentTimeMillis()) / 1000);
        RedisUtils.setEx(String.format(RedisConstants.TOKEN, vo.getToken()), expired, JSON.toJSONString(bo));
    }
```

（3）入参实体 UpdateMobileVO，对应客户端请求报文。

```java
package com.xpwu.secondary.vo;

import lombok.Getter;
import lombok.Setter;
import lombok.ToString;

/**
 * Created by IntelliJ IDEA.
 *
 * @author:caoxue
 * @date:2019/9/14 16:19
 * @description:修改手机号入参
 * @version:1.0
```

```
    */
@Setter
@Getter
@ToString(callSuper = true)
public class UpdateMobileVO extends BaseVO {
    private static final long serialVersionUID = 4656015852611448064L;

    /**
     * 新手机号
     */
    private String mobile;

    /**
     * 短信验证码
     */
    private String verifyCode;

}
```

8. 返回报文及示例

参见公共返回示例。

24.5 查询商品数量接口

1. 接口用途

此接口用于用户在个人中心模块查看我发布的、我卖出的、我买到的商品数量。

2. 参数校验

校验 token 是否为空。

3. 业务规则

此接口需要用户登录后查询，需要校验 token 有效性。

4. 请求方式

采用 GET 请求。

5. 请求报文

无。

6. 请求示例

```
{
"type":1,
"pageNum":1,
"pageSize":10
}
```

7. 编写程序

（1）ProductController 类为商品相关的接口控制层，getProductNum 方法用于个人中心模块商品数量及金额查询。主要包含校验 token 是否为空，并调用 service 层业务，将报文返回给客户端。

```
/**
 * 查询各类商品数量
 * @return
 */
@RequestMapping(value = "getProductNum", method = RequestMethod.GET)
public ResponseBO getProductNum() {
    // 校验token是否为空
    String token = checkToken();
    return ResponseBO.success(productService.getProductNum(token));
}
```

（2）ProductServiceImpl 类，getProductNum 方法用于个人中心商品数量查询业务。此接口需要登录后调用，首先校验 token 是否有效，然后根据用户编号查询，调用 Mapper 层获取数据，得到数据后重新组装。

```
/**
 * 查询各类商品数量
 * @param token
 * @return
 */
@Override
public Map<String, Object> getProductNum(String token) {
    // 校验token有效性
```

```java
        User user = userService.checkToken(token);
        ProductNumBO bo = productMapper.selectProductNum(user.getUserId());
        // 我发布的
        Map<String, Object> publish = new HashMap<>(2);
        // 我买到的
        Map<String, Object> purchase = new HashMap<>(2);
        // 我卖出的
        Map<String, Object> sale = new HashMap<>(2);
        // 我发布的商品数量与金额
        publish.put("num", bo.getPublishNum());
        publish.put("money", bo.getPublishAmount());
        // 我买到的商品数量与金额
        purchase.put("num", bo.getPurchaseNum());
        purchase.put("money", bo.getPurchaseAmount());
        // 我卖出的商品数量与金额
        sale.put("num", bo.getSaleNum());
        sale.put("money", bo.getSaleAmount());
        Map<String, Object> result = new HashMap<>(16);
        result.put("publish", publish);
        result.put("purchase", purchase);
        result.put("sale", sale);
        return result;
    }
```

（3）ProductMapper 类，自定义 selectMyProductList 接口查询。

```java
    /**
     * 个人中心——查询各类商品数量
     * @param userId
     * @return
     */
    ProductNumBO selectProductNum(@Param("userId") String userId);
```

（4）ProductMapper.xml，此处属于扩展查询，需要另外写 SQL 语句连接查询，具体如下。

```xml
    <select id="selectProductNum" parameterType="string" resultType="com.xpwu.secondary.bo.ProductNumBO">
```

```
    SELECT
        IFNULL(SUM(CASE WHEN p.publish_user_id = #{userId} AND (o.trade_status = 2 or o.order_id is null) THEN 1 ELSE 0 END), 0) publishNum,
        IFNULL(SUM(CASE WHEN p.publish_user_id = #{userId} AND (o.trade_status = 2 or o.order_id is null) THEN p.product_price ELSE 0 END), 0.00) publishAmount,
        IFNULL(SUM(CASE WHEN p.publish_user_id = #{userId} AND o.trade_status in(1, 3) AND o.selling_status = 1 THEN 1 ELSE 0 END), 0) saleNum,
        IFNULL(SUM(CASE WHEN p.publish_user_id = #{userId} AND o.trade_status in(1, 3) AND o.selling_status = 1 THEN p.product_price ELSE 0 END), 0.00) saleAmount,
        IFNULL(SUM(CASE WHEN o.buying_user_id = #{userId} AND o.trade_status in(1, 3) AND o.buying_status = 1 THEN 1 ELSE 0 END), 0) purchaseNum,
        IFNULL(SUM(CASE WHEN o.buying_user_id = #{userId} AND o.trade_status in(1, 3) AND o.buying_status = 1 THEN p.product_price ELSE 0 END), 0.00) purchaseAmount
    FROM product p
    LEFT JOIN (
      SELECT t1.* FROM 'order' t1
      INNER JOIN (
        SELECT MAX(order_id) max_order_id, o.* FROM 'order' o GROUP BY product_id
      ) t2 ON t1.order_id = t2.max_order_id
    ) o ON o.product_id = p.id

</select>
```

（5）数据层返回实体 ProductNumBO。

```
package com.xpwu.secondary.bo;

import lombok.Getter;
import lombok.Setter;
import lombok.ToString;

import java.io.Serializable;
import java.math.BigDecimal;

/**
 * Created by IntelliJ IDEA.
```

```java
 *
 * @author:caoxue
 * @date:2019/9/13 15:47
 * @description:
 * @version:1.0
 */
@Setter
@Getter
@ToString(callSuper = true)
public class ProductNumBO implements Serializable {

    private static final long serialVersionUID = 6904609746866882472L;

    /**
     * 发布商品数量
     */
    private Integer publishNum;

    /**
     * 发布商品总金额
     */
    private BigDecimal publishAmount;

    /**
     * 卖出商品数量
     */
    private Integer saleNum;

    /**
     * 卖出商品总金额
     */
    private BigDecimal saleAmount;

    /**
     * 买到商品数量
     */
    private Integer purchaseNum;

    /**
     * 买到商品总金额
     */
```

```
        private BigDecimal purchaseAmount;

}
```

8. 返回报文

查询商品数量接口报文如表 24-6 所示。

表 24-6 查询商品数量接口返回报文说明

参数名	必需	类型	字节长度	说明
publishNum	是	int	11	发布商品数量
publishAmount	是	double	15	发布商品总金额
saleNum	是	int	11	卖出商品数量
saleAmount	是	double	15	卖出商品总金额
purchaseNum	是	int	11	买到商品数量
purchaseAmount	是	double	15	买到商品总金额

9. 返回示例

```
{
"code":0,
"message":"success",
"requestId":null,
"timestamp":1567301655000,
"result":
  {
  "publish":{
    "num": 1,
    "money":100
   },
  "purchase":{
    "num": 2,
    "money":200
   },
  "sale":{
    "num": 0,
    "money":0
   }
```

        }
    }

24.6 收货地址列表接口

1. 接口用途

此接口用于客户端收货地址列表查询。

2. 参数校验

校验 token 是否为空。当前页码不传默认为 1，每页条数不传默认为 10。

3. 业务规则

此接口需要用户登录后查询，需要校验 token 有效性。

4. 请求方式

采用 POST 请求。

5. 请求报文

收货地址列表接口请求报文如表 24-7 所示。

表 24-7 收货地址列表接口请求报文说明

参数名	必需	类型	字节长度	说明
pageNum	否	int	6	当前页码，不传默认为 1
pageSize	否	int	6	每页条数，不传默认为 10

6. 请求示例

```
{
"pageNum":1,
"pageSize":10
}
```

7. 编写程序

（1）AddressController 类为收货地址相关的接口控制层，getAddressList 方法用于收货地址列表查询。主要包含校验 token 是否为空，并调用 service 层业务，将报文返回给客户端。

```java
/**
 * 获取地址列表
 * @param vo
 * @return
 */
@RequestMapping(value = "getAddressList", method = RequestMethod.POST)
public ResponseBO getAddressList(BaseVO vo) {
    // 校验token是否为空
    String token = checkToken();
    vo.setToken(token);
    // 调用service层业务
    return ResponseBO.successPageInfo(addressService.getAddressList(vo));
}
```

（2）AddressServiceImpl 类，getAddressList 方法用于收货地址列表查询业务。此接口需要登录后调用，首先校验 token 是否有效，根据用户编号查询有效数据。排序规则：优先默认地址，在此基础上按新增时间降序排列。然后开启分页查询，调用 Mapper 层获取数据。

```java
/**
 * 收货地址列表查询
 * @param vo
 * @return
 */
@Override
public List<Address> getAddressList(BaseVO vo) {
    // 校验token是否有效
    User user = userService.checkToken(vo.getToken());
    Example example = new Example(Address.class);
    // 根据用户查询有效数据
    example.createCriteria().andEqualTo("userId", user.getUserId()).andEqualTo("status", 1);
    // 按照默认地址升序（默认地址优先排列）、新增时间降序排列
    example.setOrderByClause("is_default_address asc, create_time desc");
    // 开启分页查询
    PageHelper.startPage(vo.getPageNum(), vo.getPageSize());
    return addressMapper.selectByExample(example);
}
```

（3）入参实体 BaseVO，对应客户端请求报文。

```java
package com.xpwu.secondary.vo;

import java.io.Serializable;

/**
 * Created by IntelliJ IDEA.
 *
 * @author:caoxue
 * @date:2019/8/7 14:28
 * @description:基础入参
 * @version:1.0
 */
public class BaseVO implements Serializable {

    private static final long serialVersionUID = -5146126799389106360L;
    /** 用户令牌 */
    private String token;

    /** 当前页码从 1 开始 */
    private Integer pageNum;

    /** 每页条数 */
    private Integer pageSize;

    public String getToken() {
        return token;
    }

    public void setToken(String token) {
        this.token = token;
    }

    public Integer getPageNum() {
        if (this.pageNum == null) {
            this.pageNum = 1;
        }
        return pageNum;
    }
```

```java
    public void setPageNum(Integer pageNum) {
        this.pageNum = pageNum;
    }

    public Integer getPageSize() {
        if (this.pageSize == null) {
            this.pageSize = 10;
        }
        return pageSize;
    }

    public void setPageSize(Integer pageSize) {
        this.pageSize = pageSize;
    }

    @Override
    public String toString() {
        return "BaseVO{" +
                "token='" + token + '\'' +
                ", pageNum=" + pageNum +
                ", pageSize=" + pageSize +
                '}';
    }
}
```

（4）返回报文实体 Address。

```java
package com.xpwu.secondary.entity;

import com.fasterxml.jackson.annotation.JsonIgnore;
import lombok.Getter;
import lombok.Setter;
import lombok.ToString;

import java.util.Date;
import javax.persistence.*;

@Setter
@Getter
@ToString(callSuper = true)
```

```java
@Table(name = "'address'")
public class Address {
    /**
     * 地址编号
     */
    @Id
    @Column(name = "'id'")
    @GeneratedValue(strategy = GenerationType.IDENTITY)
    private Integer id;

    /**
     * 用户编号
     */
    @Column(name = "'user_id'")
    private String userId;

    /**
     * 收货人姓名
     */
    @Column(name = "'consignee_name'")
    private String consigneeName;

    /**
     * 收货人手机号
     */
    @Column(name = "'consignee_mobile'")
    private String consigneeMobile;

    /**
     * 省
     */
    @Column(name = "'province'")
    private String province;

    /**
     * 市
     */
    @Column(name = "'city'")
    private String city;

    /**
```

```java
     * 区
     */
    @Column(name = "'district'")
    private String district;

    /**
     * 街道
     */
    @Column(name = "'street'")
    private String street;

    /**
     * 详细地址
     */
    @Column(name = "'address_detail'")
    private String addressDetail;

    /**
     * 是否为默认地址：1——是；2——否
     */
    @Column(name = "'is_default_address'")
    private Integer isDefaultAddress;

    /**
     * 新增创建时间
     */
    @Column(name = "'create_time'")
    private Date createTime;

    /**
     * 修改时间
     */
    @Column(name = "'update_time'")
    @JsonIgnore
    private Date updateTime;

    /**
     * 状态：1——有效；2——删除
     */
    @Column(name = "'status'")
    @JsonIgnore
```

```
    private Integer status;

}
```

8. 返回报文

收货地址列表接口返回报文如表 24-8 所示。

表 24-8 收货地址列表接口返回报文说明

参数名	必需	类型	字节长度	说明
id	是	int	11	地址编号
userId	是	string	32	用户编号
consigneeName	是	string	50	收货人姓名
consigneeMobile	是	string	20	收货人手机号
province	是	string	100	省
city	是	string	100	市
district	是	string	100	区
street	否	string	100	街道
addressDetail	是	string	500	详细地址
isDefaultAddress	是	int	2	是否为默认地址：1——是；2——否
createTime	是	long	13	新增时间

9. 返回示例

```
{
"code":0,
"message":"success",
"requestId":null,
"timestamp":1567301655000,
"result":[
 {
  "id":1,
  "userId":"xxx",
  "consigneeName":"张三",
  "consigneeMobile":"13888888888",
  "province":"xxx",
  "city":"xxx",
  "district":"xxx",
```

```
    "street":"xxx",
    "addressDetail":"xxx",
    "isDefaultAddress":1,
    "createTime":1567301655000
   }
  ]
 }
```

24.7 新增收货地址接口

1. 接口用途

用户新增收货地址，用于交易时选择。

2. 参数校验

校验用户令牌、收货人姓名、收货人手机号、省、市、区、街道、详细地址、是否为默认地址等参数是否为空。

3. 业务规则

校验 token 是否有效。如果当前地址为默认地址，在插入前需要先将当前数据库里面的默认地址改为非默认，即更换默认地址。

4. 请求方式

采用 POST 请求。

5. 请求报文

新增收货地址接口请求报文如表 24-9 所示。

表 24-9 新增收货地址接口请求报文说明

参数名	必需	类型	字节长度	说明
consigneeName	是	string	50	收货人姓名
consigneeMobile	是	string	20	收货人手机号
province	是	string	100	省
city	是	string	100	市
district	是	string	100	区
street	否	string	100	街道
addressDetail	是	string	500	详细地址
isDefaultAddress	是	int	2	是否为默认地址：1——是；2——否

6. 请求示例

```
{
 "consigneeName":"张三",
 "consigneeMobile":"13888888888",
 "province":"xxx",
 "city":"xxx",
 "district":"xxx",
 "street":"xxx",
 "addressDetail":"xxx",
 "isDefaultAddress":1
}
```

7. 编写程序

（1）AddressController 类，收货地址相关的接口控制层，saveAddress 方法用于用户新增收货地址。主要包含校验用户令牌、收货人姓名、收货人手机号、省、市、区、街道、详细地址、是否为默认地址等参数是否为空，然后调用 service 层业务，并返回报文信息给客户端。

```
/**
 * 新增收货地址
 * @param vo
 * @return
 */
@RequestMapping(value = "saveAddress", method = RequestMethod.POST,
produces = MediaType.APPLICATION_JSON_UTF8_VALUE)
public ResponseBO saveAddress(@RequestBody @Valid AddressVO vo) {
    // 校验token是否为空
    String token = checkToken();
    vo.setToken(token);
    addressService.saveAddress(vo);
    return ResponseBO.success();
}
```

（2）AddressServiceImpl 类，saveAddress 方法用于新增收货地址业务。首先校验 token 有效性，若校验不通过，则返回相应的错误信息。如果当前地址为默认地址，先将该用户目前的默认地址修改为非默认，然后再新增地址，保证默认地址只有一个。

```java
/**
 * 新增收货地址
 * @param vo
 */
@Override
@Transactional(rollbackFor = Exception.class, isolation = Isolation.READ_COMMITTED)
public void saveAddress(AddressVO vo) {
    // 校验token是否有效
    User user = userService.checkToken(vo.getToken());
    Address address = new Address();
    Date date = new Date();
    BeanUtils.copyProperties(vo, address);
    address.setCreateTime(date);
    address.setUpdateTime(date);
    address.setUserId(user.getUserId());
    if (1 == vo.getIsDefaultAddress()) {
        // 如果客户端传入为默认地址，将用户当前默认地址修改为非默认
        Example example = new Example(Address.class);
        example.createCriteria().andEqualTo("isDefaultAddress", 1).andEqualTo("userId", user.getUserId());
        Address update = new Address();
        update.setUpdateTime(date);
        update.setIsDefaultAddress(2);
        addressMapper.updateByExampleSelective(update, example);
    }
    // 新增收货地址
    addressMapper.insertSelective(address);
}
```

（3）入参实体 AddressVO，对应客户端请求报文。

```java
package com.xpwu.secondary.vo;

import lombok.Getter;
import lombok.Setter;
import lombok.ToString;
```

```java
import javax.validation.constraints.NotBlank;
import javax.validation.constraints.NotNull;

/**
 * Created by IntelliJ IDEA.
 *
 * @author:caoxue
 * @date:2019/10/15 22:05
 * @description:收货地址请求报文
 * @version:1.0
 */
@Setter
@Getter
@ToString(callSuper = true)
public class AddressVO extends BaseVO {
    private static final long serialVersionUID = 4995079682860629149L;

    /**
     * 地址编号
     */
    private Integer id;
    /**
     * 收货人姓名
     */
    @NotBlank(message = "收货人姓名不能为空")
    private String consigneeName;

    /**
     * 收货人手机号
     */
    @NotBlank(message = "收货人手机号不能为空")
    private String consigneeMobile;

    /**
     * 省
     */
    @NotBlank(message = "省不能为空")
    private String province;

    /**
     * 市
```

```java
     */
    @NotBlank(message = "市不能为空")
    private String city;

    /**
     * 区
     */
    @NotBlank(message = "区不能为空")
    private String district;

    /**
     * 街道
     */
    private String street;

    /**
     * 详细地址
     */
    @NotBlank(message = "详细地址不能为空")
    private String addressDetail;

    /**
     * 是否为默认地址
     */
    @NotNull(message = "是否默认地址不能为空")
    private Integer isDefaultAddress;

}
```

8. 返回报文及示例

参见公共返回示例。

24.8 修改收货地址接口

1. 接口用途

此接口用于用户修改收货地址。

2. 参数校验

校验用户令牌、地址编号是否为空。

3. 业务规则

校验 token 是否有效，校验收货地址是否存在及状态是否正常。如果当前地址为默认地址，在插入前需要先将当前数据库里面的默认地址改为非默认，即更换默认地址。

4. 请求方式

采用 POST 请求。

5. 请求报文

修改收货地址接口请求报文如表 24-10 所示。

表 24-10　修改收货地址接口请求报文说明

参数名	必需	类型	字节长度	说明
id	是	int	11	地址编号
consigneeName	否	string	50	收货人姓名
consigneeMobile	否	string	20	收货人手机号
province	否	string	100	省
city	否	string	100	市
district	否	string	100	区
street	否	string	100	街道
addressDetail	否	string	500	详细地址
isDefaultAddress	否	int	2	是否为默认地址：1——是；2——否

6. 请求示例

```
{
 "id":1,
 "consigneeName":"张三",
 "consigneeMobile":"13888888888",
 "province":"xxx",
 "city":"xxx",
 "district":"xxx",
 "street":"xxx",
 "addressDetail":"xxx",
 "isDefaultAddress":1
}
```

7. 编写程序

（1）AddressController 类，收货地址相关的接口控制层，updateAddress 方法用于用户修改收货地址。主要包含校验用户令牌、地址编号等参数是否为空，然后调用 service 层业务，并返回报文信息给客户端。

```java
/**
 * 修改收货地址
 * @param vo
 * @return
 */
@RequestMapping(value = "updateAddress", method = RequestMethod.POST)
public ResponseBO updateAddress(@RequestBody AddressVO vo) {
    // 校验token是否为空
    String token = checkToken();
    Assertion.isPositive(vo.getId(), "地址编号不能为空");
    vo.setToken(token);
    addressService.updateAddress(vo);
    return ResponseBO.success();
}
```

（2）AddressServiceImpl 类，updateAddress 方法用于修改收货地址业务。首先校验 token 有效性，若校验不通过，则返回相应的错误信息，否则根据地址编号查询地址信息是否存在及状态是否正常。如果当前传入地址为默认地址，先将该用户目前的默认地址修改为非默认，然后再修改当前地址信息。

```java
//**
 * 修改收货地址
 * @param vo
 */
@Override
@Transactional(rollbackFor = Exception.class, isolation = Isolation.READ_COMMITTED)
public void updateAddress(AddressVO vo) {
    // 校验token是否有效
    userService.checkToken(vo.getToken());
    Address address = addressMapper.selectByPrimaryKey(vo.getId());
    Assertion.notNull(address, "收货地址不存在");
```

```java
        Assertion.isTrue(1 == address.getStatus(), "收货地址已被删除");
        Address update = new Address();
        Date date = new Date();
// 将请求报文中传入的需要修改的字段复制到 update 对象中
        BeanUtils.copyProperties(vo, update);
        update.setUpdateTime(date);
        if (Detect.isPositive(vo.getIsDefaultAddress()) && 1 == vo.getIsDefaultAddress()) {
            // 如果客户端传入为默认地址，将用户当前默认地址修改为非默认
            Example example = new Example(Address.class);
            example.createCriteria().andEqualTo("isDefaultAddress", 1).andNotEqualTo("id", vo.getId());
            Address entity = new Address();
            entity.setUpdateTime(date);
            entity.setIsDefaultAddress(2);
            addressMapper.updateByExampleSelective(entity, example);
        }
        addressMapper.updateByPrimaryKeySelective(update);
    }
```

（3）入参实体 AddressVO，对应客户端请求报文。

```java
package com.xpwu.secondary.vo;

import lombok.Getter;
import lombok.Setter;
import lombok.ToString;

import javax.validation.constraints.NotBlank;
import javax.validation.constraints.NotNull;

/**
 * Created by IntelliJ IDEA.
 *
 * @author:caoxue
 * @date:2019/10/15 22:05
 * @description:收货地址请求报文
 * @version:1.0
 */
@Setter
```

```java
@Getter
@ToString(callSuper = true)
public class AddressVO extends BaseVO {
    private static final long serialVersionUID = 4995079682860629149L;

    /**
     * 地址编号
     */
    private Integer id;
    /**
     * 收货人姓名
     */
    @NotBlank(message = "收货人姓名不能为空")
    private String consigneeName;

    /**
     * 收货人手机号
     */
    @NotBlank(message = "收货人手机号不能为空")
    private String consigneeMobile;

    /**
     * 省
     */
    @NotBlank(message = "省不能为空")
    private String province;

    /**
     * 市
     */
    @NotBlank(message = "市不能为空")
    private String city;

    /**
     * 区
     */
    @NotBlank(message = "区不能为空")
    private String district;

    /**
     * 街道
```

```
     */
    private String street;

    /**
     * 详细地址
     */
    @NotBlank(message = "详细地址不能为空")
    private String addressDetail;

    /**
     * 是否为默认地址
     */
    @NotNull(message = "是否为默认地址不能为空")
    private Integer isDefaultAddress;

}
```

8. 返回报文及示例

参见公共返回示例。

24.9 删除收货地址接口

1. 接口用途

此接口用于用户删除收货地址。

2. 参数校验

校验用户令牌、地址编号是否为空。

3. 业务规则

校验 token 是否有效,校验收货地址是否存在及状态是否正常。如果当前地址为默认地址,在插入前需要先将当前数据库里面的默认地址改为非默认,即更换默认地址。

4. 请求方式

采用 POST 请求。

5. 请求报文

修改收货地址接口请求报文如表 24-11 所示。

表 24-11　修改收货地址接口请求报文说明

参数名	必需	类型	字节长度	说明
id	是	int	11	地址编号
consigneeName	否	string	50	收货人姓名
consigneeMobile	否	string	20	收货人手机号
province	否	string	100	省
city	否	string	100	市
district	否	string	100	区
street	否	string	100	街道
addressDetail	否	string	500	详细地址
isDefaultAddress	否	int	2	是否为默认地址：1——是；2——否

6. 请求示例

```
{
 "id":1,
 "consigneeName":" 张三 ",
 "consigneeMobile":"13888888888",
 "province":"xxx",
 "city":"xxx",
 "district":"xxx",
 "street":"xxx",
 "addressDetail":"xxx",
 "isDefaultAddress":1
}
```

7. 编写程序

（1）AddressController 类，收货地址相关的接口控制层，delAddress 方法用于用户删除收货地址。主要包含校验用户令牌、地址编号等参数是否为空，然后调用 service 层业务，并返回报文信息给客户端。

```
/**
 * 删除收货地址
 * @param vo
 * @return
 */
@RequestMapping(value = "delAddress", method = RequestMethod.POST)
```

```java
public ResponseBO delAddress(@RequestBody AddressVO vo) {
    // 校验token是否为空
    String token = checkToken();
    Assertion.isPositive(vo.getId(), "地址编号不能为空");
    vo.setToken(token);
    addressService.delAddress(vo);
    return ResponseBO.success();
}
```

（2）AddressServiceImpl 类，delAddress 方法用于删除收货地址业务。首先校验 token 有效性，若校验不通过，则返回相应的错误信息，否则根据地址编号查询地址信息是否存在及状态是否正常。然后对地址信息进行逻辑删除，即将 status 修改为 2，完成删除操作。

```java
/**
 * 删除收货地址
 * @param vo
 */
@Override
public void delAddress(AddressVO vo) {
    // 校验token是否有效
    userService.checkToken(vo.getToken());
    Address address = addressMapper.selectByPrimaryKey(vo.getId());
    Assertion.notNull(address, "收货地址不存在");
    Assertion.isTrue(1 == address.getStatus(), "收货地址已被删除，请勿重复操作");
    Address update = new Address();
    // 执行逻辑删除
    update.setStatus(2);
    update.setUpdateTime(new Date());
    update.setId(vo.getId());
    addressMapper.updateByPrimaryKeySelective(update);
}
```

（3）入参实体 AddressVO，对应客户端请求报文。此处共用新增与修改接口的入参实体，省略了其他属性。

```
package com.xpwu.secondary.vo;
```

```java
import lombok.Getter;
import lombok.Setter;
import lombok.ToString;

import javax.validation.constraints.NotBlank;
import javax.validation.constraints.NotNull;

/**
 * Created by IntelliJ IDEA.
 *
 * @author:caoxue
 * @date:2019/10/15 22:05
 * @description:收货地址请求报文
 * @version:1.0
 */
@Setter
@Getter
@ToString(callSuper = true)
public class AddressVO extends BaseVO {
    private static final long serialVersionUID = 4995079682860629149L;

    /**
     * 地址编号
     */
    private Integer id;
}
```

8. 返回报文及示例

参见公共返回示例。

第 25 章 后端环境部署

【本章导读】

◎ 项目部署流程

项目部署流程

后端项目部署流程主要分为以下 3 个步骤：配置编译环境，打包并上传，启动服务。

1. 配置编译环境

正常的项目开发会涉及多个环境，后端主要会有 dev 开发环境、prod 生产环境、test 测试环境。配置不同的编译环境也有很多种方法，但中心思想还是通过参数控制打包不同的环境，这里通过不同文件夹区分不同的环境，具体配置如下。

（1）resources 资源文件配置，通过在 profiles 下建立不同的文件夹来实现，如图 25-1 所示。

图 25-1

（2）pom.xml 配置打包环境参数。

```
<profiles>
```

```xml
<profile>
    <id>dev</id>
    <properties>
        <package.environment>profiles/dev</package.environment>
    </properties>
</profile>
<profile>
    <id>prod</id>
    <properties>
        <package.environment>profiles/prod</package.environment>
    </properties>
</profile>
</profiles>
<profile>
    <id>test</id>
    <properties>
        <package.environment>profiles/test</package.environment>
    </properties>
</profile>
```

并且在 <build></build> 标签中加入如下打包配置。

```xml
<resources>
    <resource>
        <directory>src/main/resources</directory>
        <filtering>false</filtering>
        <excludes>
            <exclude>profiles/**</exclude>
        </excludes>
    </resource>
    <resource>
        <directory>src/main/resources/${package.environment}</directory>
        <filtering>true</filtering>
        <excludes>
            <exclude>profiles/**</exclude>
        </excludes>
    </resource>
</resources>
```

这样配置完成之后，就可以勾选对应的环境选项进行打包，如图 25-2 所示。也可以通过 Maven 命令 mvn clean package -P dev 进行打包。

图 25-2

2．打包并上传

（1）打包完成之后，在项目目录下会生成一个 target 目录，该目录下有一个 Jar 包，如图 25-3 所示。

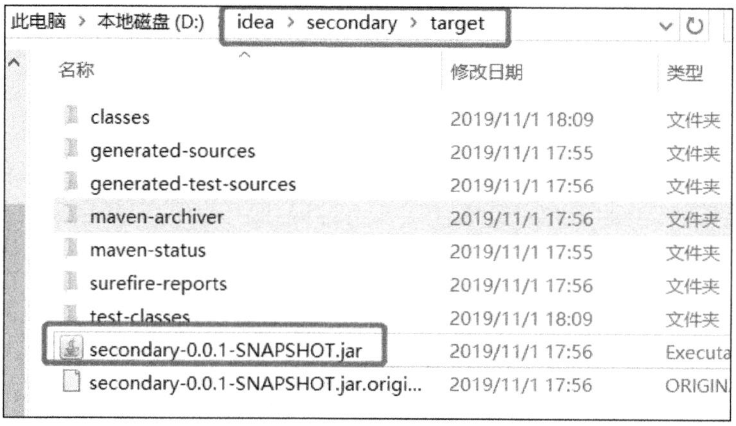

图 25-3

（2）上传 Jar 包至服务器。通过命令 mkdir -p home/secondary 创建文件路径 /home/secondary，紧接着进入该目录 cd /home/secondary，通过命令 rz 上传 Jar 包。

3．启动服务

Jar 包上传成功之后就可以启动项目，通过如下命令即可启动。

nohup java -jar -Xmx1024M secondary-0.0.1-SNAPSHOT.jar >/dev/null 2>&1 &

nohup 表示不挂断的运行命令，不会随着 Ctrl+C 退出后导致项目停止。

-Xmx1024M 是设置 JVM 最大可用内存为 1024MB。

>/dev/null 表示输出重定向到 null，即不输出到控制台。

2>&1 表示将错误信息定位到输出。

& 表示后台运行，一般跟 nohup 配合使用。

至此，后端项目部署就完成了。

第 26 章 后端开发总结

【本章导读】
◎ 开发思路总结
◎ 开发难点总结

26.1 开发思路总结

本次项目开发，有以下几个步骤及注意点。

（1）理解需求，一定要多分析商品原型，明白我们要开发的是一个什么样的项目，可以画一些辅助的流程图，帮助自己理解。

（2）数据建模，也就是表结构设计，这一步很关键。任何项目底层都需要依赖于各种数据，建立正确的数据流与数据结构可以很大程度上减小开发难度。

（3）项目搭建，技术选型上也要慎重，不要一味地追求新技术，刚出来的东西不一定就是最好的，要选一些稳定的框架及版本。然后就是基础的组件封装，比如说业务层可以封装一些基础的增删改查，减少重复的接口开发，也可以进行异常的统一处理，这样代码看着比较简洁，基本上不需要 try.... catch。

（4）项目开发，作为一个有追求的开发者来说，编码规范是一定要注意的。推荐大家用阿里巴巴的编码规范，坚持使用，能够提高自己的编码能力及改善自己的编码习惯。

26.2 开发难点总结

本次开发过程中有以下几个难点，在此分享给大家。
1. 难点一
使用 websocket 长连接实现用户之间的即时通信，也就是在线聊天，在 22.4 节中也详细讲了接入流程。

2. 难点二

订单超时自动取消业务。前面的章节中没有对这块做详细讲解，在这里跟大家讲解一下。

刚开始想用定时任务，每几分钟轮询一次，看订单是否超时，不过这个想法很快就被否定了。这种方式比较消耗性能，而且这个频率也不好控制，执行频率高了（比如每2/5/10秒1次），服务器"吃不消"，执行频率低了（比如每2/5/10分钟1次），数据都是被延时处理的。

举个例子，假设每5分钟调度一次，用户在当天10：21下单，那么到第二天10：20的时候，调度任务扫描不到这个订单（还未超时），到了10：25的时候，这笔订单才被扫描到，并自动取消，体验不好。

综上考虑，使用RabbitMQ的延时队列实现此需求比较合适。实现原理是先发送一条消息到死信队列上并设置有效时间，消息超时之后会通过死信路由转发到正常消费队列中，这样间接中转实现了延时队列，具体实现如下。

（1）在pom文件中添加amqp引用。

```xml
<!-- RabbitMQ -->
<dependency>
    <groupId>org.springframework.boot</groupId>
    <artifactId>spring-boot-starter-amqp</artifactId>
</dependency>
```

（2）在application.properties文件中添加如下配置。

```
#RabbitMQ 配置
## 指定 RabbitMQ host
spring.rabbitmq.host = 127.0.0.1
## 端口
spring.rabbitmq.port = 5672
## 虚拟机
spring.rabbitmq.virtualHost = /secondary-host
## 用户名
spring.rabbitmq.username = admin
## 密码
```

```
spring.rabbitmq.password = 123456
## 指定最大的消费者数量
spring.rabbitmq.maxConcurrentConsumers = 5
## 指定最小的消费者数量
spring.rabbitmq.concurrentConsumers = 1
## 设置超时时间
spring.rabbitmq.connection-timeout = 5000
## 手动确认
spring.rabbitmq.listener.simple.acknowledge-mode=manual
## 订单队列延时时间，本地2min
spring.rabbitmq.order.ttl=120000
```

（3）新建 RabbitMQ 配置。

```
package com.xpwu.secondary.rabbitmq.config;

import org.springframework.amqp.rabbit.config.SimpleRabbitListenerContainerFactory;
import org.springframework.amqp.rabbit.connection.CachingConnectionFactory;
import org.springframework.amqp.rabbit.connection.ConnectionFactory;
import org.springframework.amqp.rabbit.core.RabbitAdmin;
import org.springframework.amqp.rabbit.core.RabbitTemplate;
import org.springframework.beans.factory.annotation.Qualifier;
import org.springframework.beans.factory.annotation.Value;
import org.springframework.boot.autoconfigure.amqp.SimpleRabbitListenerContainerFactoryConfigurer;
import org.springframework.context.annotation.Bean;
import org.springframework.context.annotation.Configuration;
import org.springframework.context.annotation.Primary;

/**
 * @author caoxue
 * @description rabbitmq 配置
 */
@Configuration
public class RabbitMqConfig {

    /**
     * 连接工厂
     * @param host
```

```java
 * @param port
 * @param username
 * @param password
 * @param virtualHost
 * @return
 */
@Primary
@Bean(name = "connectionFactory")
public ConnectionFactory connectionFactory(@Value("${spring.rabbitmq.host}") String host,
                                           @Value("${spring.rabbitmq.port}") int port,
                                           @Value("${spring.rabbitmq.username}") String username,
                                           @Value("${spring.rabbitmq.password}") String password,
                                           @Value("${spring.rabbitmq.virtualHost}") String virtualHost) {
    CachingConnectionFactory connectionFactory = new CachingConnectionFactory();
    connectionFactory.setHost(host);
    connectionFactory.setPort(port);
    connectionFactory.setUsername(username);
    connectionFactory.setPassword(password);
    connectionFactory.setVirtualHost(virtualHost);
    return connectionFactory;
}

@Primary
@Bean(name = "rabbitTemplate")
public RabbitTemplate rabbitTemplate(@Qualifier("connectionFactory") ConnectionFactory connectionFactory) {
    return new RabbitTemplate(connectionFactory);
}

@Bean(name = "rabbitAdmin")
RabbitAdmin rabbitAdmin(@Qualifier("connectionFactory") ConnectionFactory connectionFactory) {
    return new RabbitAdmin(connectionFactory);
}
```

```
    /**
     * 使用@RabbitListener注解监听所需连接工厂 SimpleRabbitListenerContainerFactory
     * @param configurer
     * @param connectionFactory
     * @param maxConcurrentConsumers
     * @param concurrentConsumers
     * @return
     */
    @Bean(name = "commomFactory")
    public SimpleRabbitListenerContainerFactory commonFactory(SimpleRabbitListenerContainerFactoryConfigurer configurer,
            @Qualifier("connectionFactory") ConnectionFactory connectionFactory,
            @Value("${spring.rabbitmq.maxConcurrentConsumers}") int maxConcurrentConsumers,
            @Value("${spring.rabbitmq.concurrentConsumers}") int concurrentConsumers) {
        SimpleRabbitListenerContainerFactory factory = new SimpleRabbitListenerContainerFactory();
        factory.setMaxConcurrentConsumers(maxConcurrentConsumers);
        factory.setConcurrentConsumers(concurrentConsumers);
        configurer.configure(factory, connectionFactory);
        return factory;
    }
}
```

（4）新建枚举类，维护队列名称、路由、交换机。

```
package com.xpwu.secondary.enums;

/**
 * 定义RabbitMQ需要的常量
 * @author caoxue
 */
public class RabbitMqEnum {

    /**
     * 队列名称
     */
```

```java
            public enum QueueName {
                /**
                 * 队列名称枚举
                 */
                SECONDARY_ORDER_QUEUE("SECONDARY_ORDER_QUEUE", "订单转发队列"),
                SECONDARY_ORDER_DEAD_LETTER_QUEUE("SECONDARY_ORDER_DEAD_LETTER_QUEUE", "订单死信队列"),
                ;

                private String code;
                private String name;

                QueueName(String code, String name) {
                    this.code = code;
                    this.name = name;
                }

                public String getCode() {
                    return code;
                }

                public String getName() {
                    return name;
                }

            }

            /**
             * describe: 定义 routing_key
             * creat_user:admin
             * creat_date:2017/10/31
             **/
            public enum RoutingEnum {
                /**
                 * 路由名称枚举
                 */
                SECONDARY_ORDER_ROUTING("SECONDARY_ORDER_ROUTING", "订单转发路由"),
                SECONDARY_ORDER_DEAD_LETTER_ROUTING("SECONDARY_ORDER_DEAD_LETTER_ROUTING", "订单死信路由"),
                ;
```

```java
        private String code;
        private String name;

        RoutingEnum(String code, String name) {
            this.code = code;
            this.name = name;
        }

        public String getCode() {
            return code;
        }

        public String getName() {
            return name;
        }
    }
    /**
     * @Description:定义数据交换方式
     * @author admin
     */
    public enum Exchange {
        /**
         * 交换机名称枚举
         */
        SECONDARY_ORDER_DEAD_LETTER_ECHANGE("SECONDARY_ORDER_DEAD_LETTER_ECHANGE", "订单死信交换机"),
        ;

        private String code;
        private String name;

        Exchange(String code, String name) {
            this.code = code;
            this.name = name;
        }

        public String getCode() {
            return code;
        }

        public String getName() {
```

```
                return name;
        }
    }
}
```

（5）新建常量类，定义死信队列交换机及路由 key。

```
package com.xpwu.secondary.constants;

/**
 * Created by IntelliJ IDEA.
 *
 * @author:caoxue
 * @date:2019/10/15 9:27
 * @description:rabbitMq 常量
 * @version:1.0
 */
public class RabbitMqConstants {

    /**
     * 死信队列交换机
     */
    public final static String DEAD_LETTER_EXCHANGE = "x-dead-letter-exchange";

    /**
     * 死信队列路由 key
     */
    public final static String DEAD_LETTER_ROUTING_KEY = "x-dead-letter-routing-key";
}
```

（6）新建 MqBindConfig 类，用于队列与交换机的绑定。

```
package com.xpwu.secondary.rabbitmq.config.binding;

import com.xpwu.secondary.enums.RabbitMqEnum;
import com.xpwu.secondary.rabbitmq.config.RabbitMqConfig;
import lombok.extern.slf4j.Slf4j;
```

```java
import org.slf4j.Logger;
import org.slf4j.LoggerFactory;
import org.springframework.amqp.core.*;
import org.springframework.amqp.rabbit.core.RabbitAdmin;
import org.springframework.beans.factory.annotation.Qualifier;
import org.springframework.context.annotation.Bean;
import org.springframework.context.annotation.Configuration;

/**
 * 用于配置交换机和队列对应关系
 * 新增消息队列应该按照如下步骤
 * (1) 增加 queue bean，参见 queueXXXX 方法
 * (2) 增加 queue 和 exchange 的 binding
 *
 * @author caoxue
 **/
@Configuration
@Slf4j
public class MqBindingConfig extends RabbitMqConfig {

    /**
     * 订单队列绑定死信交换机
     * @param orderQueue
     * @param orderDeadLetterExchange
     * @param rabbitAdmin
     * @return
     */
    @Bean
    Binding bindingOrderQueue(@Qualifier("orderQueue") Queue orderQueue,
                              @Qualifier("orderDeadLetterExchange") DirectExchange orderDeadLetterExchange,
                              RabbitAdmin rabbitAdmin){
        Binding binding = BindingBuilder.bind(orderQueue)
                    .to(orderDeadLetterExchange).with(RabbitMqEnum
                    .RoutingEnum.SECONDARY_ORDER_ROUTING.getCode());
        rabbitAdmin.declareBinding(binding);
        log.debug(" 订单队列绑定 orderDeadLetterExchange 交换机成功 ");
        return binding;
    }

    /**
```

```
     * 死信队列绑定死信交换机
     * @param orderDeadLetterQueue
     * @param orderDeadLetterExchange
     * @param rabbitAdmin
     * @return
     */
    @Bean
    public Binding bindingOrderDeadLetterQueue(@Qualifier("orderDead
LetterQueue") Queue orderDeadLetterQueue,
                                                @Qualifier("orderDeadLette
rExchange") DirectExchange orderDeadLetterExchange,
                                                RabbitAdmin rabbitAdmin) {
        Binding binding = BindingBuilder.bind(orderDeadLetterQueue)
                .to(orderDeadLetterExchange).with(RabbitMqEnum
                        .RoutingEnum.SECONDARY_ORDER_DEAD_LETTER_
ROUTING.getCode());
        rabbitAdmin.declareBinding(binding);
        log.debug("订单死信队列绑定orderDeadLetterExchange交换机成功");
        return binding;
    }
}
```

（7）新建MqExchangeConfig类，定义交换机。

```
package com.xpwu.secondary.rabbitmq.config.exchange;

import com.xpwu.secondary.enums.RabbitMqEnum;
import com.xpwu.secondary.rabbitmq.config.RabbitMqConfig;
import lombok.extern.slf4j.Slf4j;
import org.slf4j.Logger;
import org.slf4j.LoggerFactory;
import org.springframework.amqp.core.DirectExchange;
import org.springframework.amqp.rabbit.core.RabbitAdmin;
import org.springframework.context.annotation.Bean;
import org.springframework.context.annotation.Configuration;

/**
 * 用于配置交换机和队列对应关系
 * 新增消息队列应该按照如下步骤
 * （1）增加queue bean，参见queueXXXX方法
 * （2）增加queue和exchange的binding
```

```
 *
 * @author caoxue
 **/
@Configuration
@Slf4j
public class MqExchangeConfig extends RabbitMqConfig {
    /**
     * 定义死信交换机
     * @param rabbitAdmin
     * @return
     */
    @Bean("orderDeadLetterExchange")
    DirectExchange orderDeadLetterExchange(RabbitAdmin rabbitAdmin) {
        DirectExchange orderDeadLetterExchange = new DirectExchange(RabbitMqEnum.Exchange.SECONDARY_ORDER_DEAD_LETTER_ECHANGE.getCode());
        rabbitAdmin.declareExchange(orderDeadLetterExchange);
        log.info("死信交换机-点对点初始化成功!");
        return orderDeadLetterExchange;
    }
}
```

（8）新建 DirectQueueConfig 类，定义队列。

```
package com.xpwu.secondary.rabbitmq.config.queue;

import com.xpwu.secondary.constants.RabbitMqConstants;
import com.xpwu.secondary.enums.RabbitMqEnum;
import com.xpwu.secondary.rabbitmq.config.RabbitMqConfig;
import lombok.extern.slf4j.Slf4j;
import org.springframework.amqp.core.Queue;
import org.springframework.amqp.rabbit.core.RabbitAdmin;
import org.springframework.context.annotation.Bean;
import org.springframework.context.annotation.Configuration;

import java.util.HashMap;
import java.util.Map;

/**
 * 增加新的队列
 * @author caoxue
 */
```

```java
@Configuration
@Slf4j
public class DirectQueueConfig extends RabbitMqConfig {
    /**
     * 订单死信队列
     * @param rabbitAdmin
     * @return
     */
    @Bean("orderDeadLetterQueue")
    Queue orderDeadLetterQueue(RabbitAdmin rabbitAdmin){
        Map<String, Object> args = new HashMap<>(16);
        args.put(RabbitMqConstants.DEAD_LETTER_EXCHANGE, RabbitMqEnum.Exchange.SECONDARY_ORDER_DEAD_LETTER_ECHANGE.getCode());
        // 通过订单路由key 将死信队列路由到订单转发队列，所以消费端只需要监听订单转发队列就可以了
        args.put(RabbitMqConstants.DEAD_LETTER_ROUTING_KEY, RabbitMqEnum.RoutingEnum.SECONDARY_ORDER_ROUTING.getCode());
        Queue queue = new Queue(RabbitMqEnum.QueueName.SECONDARY_ORDER_DEAD_LETTER_QUEUE.getCode() ,true, false, false, args);
        rabbitAdmin.declareQueue(queue);
        log.info("queue {} create successed" ,RabbitMqEnum.QueueName.SECONDARY_ORDER_DEAD_LETTER_QUEUE.getCode());
        return queue;
    }

    /**
     * 订单转发队列
     * @param rabbitAdmin
     * @return
     */
    @Bean("orderQueue")
    Queue orderQueue(RabbitAdmin rabbitAdmin){
        Queue queue = new Queue(RabbitMqEnum.QueueName.SECONDARY_ORDER_QUEUE.getCode() ,true);
        rabbitAdmin.declareQueue(queue);
        log.info("queue {} create successed" ,RabbitMqEnum.QueueName.SECONDARY_ORDER_QUEUE.getCode());
        return queue;
    }

}
```

（9）基础配置已经完成，接下来新建 DirectMqSender 类，用于发送消息。

```java
package com.xpwu.secondary.rabbitmq.producer;

import com.xpwu.secondary.enums.RabbitMqEnum;
import lombok.extern.slf4j.Slf4j;
import org.springframework.amqp.core.MessageDeliveryMode;
import org.springframework.amqp.core.MessageProperties;
import org.springframework.amqp.rabbit.connection.CorrelationData;
import org.springframework.amqp.rabbit.core.RabbitTemplate;
import org.springframework.beans.factory.annotation.Autowired;
import org.springframework.beans.factory.annotation.Value;
import org.springframework.stereotype.Component;

import javax.annotation.Resource;
import java.util.UUID;

/**
 * @author caoxue
 * rabbitmq 发送消息工具类
 */
@Component
@Slf4j
public class DirectMqSender implements RabbitTemplate.ConfirmCallback {

    @Value("${spring.rabbitmq.order.ttl}")
    private int ttl;

    @Resource(name = "rabbitTemplate")
    private RabbitTemplate rabbitTemplate;

    @Autowired
    public DirectMqSender(RabbitTemplate rabbitTemplate) {
        this.rabbitTemplate = rabbitTemplate;
        this.rabbitTemplate.setConfirmCallback(this);
    }

    @Override
    public void confirm(CorrelationData correlationData, boolean b, String s) {
        log.info("confirm:" + correlationData.getId());
```

        }

        /**
         * 发送延时消息
         * @param route
         * @param orderId
         */
        public void sendTtlMessage(RabbitMqEnum.RoutingEnum route, String orderId) {
            CorrelationData correlationData = new CorrelationData(UUID.randomUUID().toString());
            log.info("send ttl:" + correlationData.getId());
            log.info("ttl:" + ttl);
            this.rabbitTemplate.convertAndSend(RabbitMqEnum.Exchange.SECONDARY_ORDER_DEAD_LETTER_ECHANGE.getCode(), route.getCode(), orderId, message -> {
                MessageProperties messageProperties = message.getMessageProperties();
                // 设置消息使用持久模式，可以不设置，默认也是持久的
                messageProperties.setDeliveryMode(MessageDeliveryMode.PERSISTENT);
                // 设置编码格式
                messageProperties.setContentEncoding("UTF-8");
                // 设置延时时间，单位为 ms
                messageProperties.setExpiration(ttl + "");
                return message;
            }, correlationData);
        }
    }

    // 以下是发送调用代码，往死信队列上发送延时消息
    // 发送延时队列，用于 24 小时订单超时自动取消
    directMqSender.sendTtlMessage(RabbitMqEnum.RoutingEnum.SECONDARY_ORDER_DEAD_LETTER_ROUTING, orderId + "");
```

（10）最后就是对消息进行监听了，新建一个 OrderConsumer 类，监听订单转发队列，当死信消息超时，通过路由转发到订单队列进行消费。至此，订单超时自动取消的功能就完成了。

```
    package com.xpwu.secondary.rabbitmq.consumer;

```java
import com.rabbitmq.client.Channel;
import com.xpwu.secondary.entity.Order;
import com.xpwu.secondary.mapper.OrderMapper;
import com.xpwu.secondary.utils.Assertion;
import com.xpwu.secondary.utils.JacksonUtils;
import com.xpwu.secondary.vo.CommonOrderVO;
import lombok.extern.slf4j.Slf4j;
import org.springframework.amqp.core.Message;
import org.springframework.amqp.rabbit.annotation.RabbitListener;
import org.springframework.beans.factory.annotation.Autowired;
import org.springframework.context.annotation.Configuration;

import java.io.IOException;
import java.nio.charset.Charset;
import java.util.Date;

/**
 * Created by IntelliJ IDEA.
 *
 * @author:caoxue
 * @date:2019/10/14 21:34
 * @description:订单监听队列
 * @version:1.0
 */
@Configuration
@Slf4j
public class OrderConsumer {

    @Autowired
    private OrderMapper orderMapper;

    /**
     * 监听订单转发队列 ，接收超时的死信队列消息
     * @param message
     * @param channel
     * @throws IOException
     */
    @RabbitListener(queues = "SECONDARY_ORDER_QUEUE", containerFactory = "commomFactory")
    public void onMessage(Message message, Channel channel) throws IOException {
```

```java
            try {
                String body = new String(message.getBody(), Charset.forName("UTF-8"));
                log.info("orderConsumer is ok! 传入参数: " + body);
                CommonOrderVO bo = (CommonOrderVO) JacksonUtils.jsonToBean(body, CommonOrderVO.class);
                // 超时，查询订单状态，若已完成，则不处理，反之自动取消
                Order order = orderMapper.selectByPrimaryKey(bo.getOrderId());
                Assertion.notNull(order, " 订单信息不存在 ");
                // 订单还未取消或未结算，自动取消
                boolean trade = order.getPayStatus() == 1 || order.getPayStatus() == 4;
                if (order.getTradeStatus() == 1 && trade) {
                    Order update = new Order();
                    update.setOrderId(order.getOrderId());
                    update.setTradeStatus(2);
                    update.setUpdateTime(new Date());
                    orderMapper.updateByPrimaryKeySelective(update);
                }
            } catch (Exception e) {
                log.error("orderConsumer is Exception! 错误信息:", e);
            } finally {
                // 手动应答
                channel.basicAck(message.getMessageProperties().getDeliveryTag(), false);
            }
        }
    }
```

3. 难点三

支付宝 WAP 支付对接，在 23.4 节中也介绍了支付宝接入流程。其实支付宝对接本身并不难，把交互流程弄清楚即可。在这里就不再过多介绍了，希望前面的流程能够帮助大家快速上手支付宝支付接口对接。